BLACK GOLD IN CALIFORNIA

The Story of the California Petroleum Industry
by Robert D. Francis, Ph.D.

A publication of the
California Independent Petroleum Association

HPNbooks
A division of Lammert Incorporated
San Antonio, Texas

DEDICATION

CIPA would like to dedicate this book to Phil Ryall, retired petroleum geologist and historian, who helped in the research of this project.

 Phil Ryall with Daphne Fletcher at the CIPA Golf Outing held at the Bakersfield Country Club in November 2014. Mr. Ryall had many good friends in the oil industry!

First Edition

Copyright © 2016 HPNbooks

ISBN: 978-1-944891-13-8

Library of Congress Card Catalog Number: 2016943636

Black Gold in California: The Story of the California Petroleum Industry

author:	Robert D. Francis, Ph.D.
cover photographer:	Greg Iger
contributing writer for "Sharing the Heritage":	Joe Goodpasture

HPNbooks

president:	Ron Lammert
project manager:	Daphne Fletcher
assistant project manager:	Anita Andersen
administration:	Donna M. Mata, Melissa G. Quinn
book sales:	Dee Steidle
production:	Colin Hart, Evelyn Hart, Glenda Tarazon Krouse
	Tim Lippard, Christopher D. Sturdevant

The California Independent Petroleum Association would like to thank the Chevron Corporation for donating pictures used in this publication.

PRINTED IN SOUTH KOREA

CONTENTS

Union Oil Company of California, unknown date.

PHOTOGRAPH COURTESY OF THE PETRU CORPORATION.

Howard Supply Company
3824 Buck Owens Boulevard
Bakersfield, California 93308
661-324-9721
www.howard-supply.com

California Resources Corporation
9200 Oakdale Avenue
Los Angeles, California 91311
818-661-6000
www.crc.com

LEGACY SPONSORS

Through their generous support, these companies helped to make this project possible.

Cavins Oil Well Tools
2598 East 28th Street
Signal Hill, California 90755
562-424-8564
www.cavins.com

JD Rush Company
5900 East Lerdo Highway
Shafter, California 93263
661-392-1900
www.jdrush.com

Petrol Transport, Inc.
5502 South Granite Road
Bakersfield, California 93308
661-393-6514
www.petroltransportinc.com

PROLOGUE

The history of petroleum in California begins 10 to 18 million years ago, during the Miocene Epoch, in a land lost in time. Volcanoes, long since vanished, laid down ash deposits a mile thick. Our deserts were then grassland and forest. Camels, rhinos, flamingoes, and little *Merychippus*, the three-foot-tall ancestor of the modern horse, roamed the land. The volcanoes flanked great ocean basins that deepened and acted as storehouses for the Monterey Formation (Shale), source of nearly all the petroleum in California. Eventually these basins filled up with sediments; some became land in places that were later given names like San Joaquin, Ventura, Santa Maria, and Los Angeles. Slowly, organic matter in the Monterey Formation was released to form oil.

These basins may be small geographically, and other places on earth may have yielded more petroleum, but California knows no equal when it comes to concentrated richness of deposits close to the surface. What is the most prolific province in terms of barrels of oil produced per cubic mile of oil-bearing strata? Saudi Arabia? The North Sea? The Gulf of Mexico? No; it is Los Angeles Basin.

Humans arrived in California, first Indians and then Europeans. They farmed the rich land, built towns and made lives for themselves. At first the incalculable treasure beneath their feet was unknown to them, although it manifested itself at the surface in various places as tar seeps or even rivers of asphalt. By 1903 California became the largest producing state in the United States, peaking at over 1 million barrels a day in the 1970s. It is still producing over half a million a day, making it the number three state. This great resource would never have been utilized for the benefit of mankind had it not been for the pioneers of the early petroleum industry. The risk-taking, hard work, and innovating spirit of those entrepreneurs are still alive today in companies great and small, which even now are finding new ways to unlock America's hidden bounty. This is a story about people—past and present—who made the mighty energy industry of today possible.

Carbonaceous marl unit of the Monterey Formation (or Shale), Gaviota Beach, California. This rock can contain as much as twenty-three percent organic carbon, which is the source of crude oil. The layers dip steeply to the left in the picture, and are seen both in the cliff face, and on the beach where they have been eroded by the waves. The formation continues offshore, where it dips to great depths under the ocean. The ocean in this picture (left background) is at low tide. Pressure and heat at great depth turned the organic matter into the crude oil found in many oil fields in coastal and offshore Santa Barbara County.

COURTESY OF PROFESSOR RICHARD BEHL, CALIFORNIA STATE UNIVERSITY LONG BEACH.

CHAPTER ONE

DAWN: 1850-1900

GOLD AND ASPHALT

A time of transition. Maricopa (San Joaquin Valley) rail siding with tank cars and oil derricks in the background. Mule-drawn freight wagons are being used to carry pipe to where it is needed.
COURTESY OF CALIFORNIA OIL MUSEUM, SANTA PAULA, CALIFORNIA.

In the 1850s the gold rush was already waning in the infant State of California, although immigrants continued to flow in for decades, attracted by the fertile land and mild climate. A major problem in those early years of statehood was isolation from the rest of the Union, even after the opening of the transcontinental railroad in 1869. Coal was America's main energy source for factories and transportation. California had to import its coal from the East, and high cost inhibited the development of industry in the state. The discovery of California's own oil changed all that. Before the century was out, the state's railroads were using petroleum in their locomotives, factories were using it in their furnaces and stationary engines, and mines in their motive equipment. The industrial revolution in California was on. By 1900 everything was new. California stood on the threshold of a dynamic twentieth century of industry, fueled by petroleum. This chapter is the story of this remarkable transformation, and the nineteenth century people who made it happen.

For uncounted centuries native people harvested "asphaltum" from seeps and tar pits. They used it to seal baskets, attach arrowheads to their shafts, and as glue to make brushes to help grind acorn meal. The Tongva people, who inhabited Santa Catalina Island for up to 7,000 years, probably used bitumen to caulk canoes in which they made the twenty-five-mile journey to the mainland to trade with other tribes. Spaniards used the tar as roofing material in missions and other buildings. "Brea," or tar, is used today in place names all over the state, perhaps most famously at the La Brea Tar pits, a natural oil seep in the Los Angeles Basin that is still active today. For the immense Spanish-grant rancheros, mainstays of the early California beef economy, the tar seeps and pits were a nuisance, as they trapped and killed numerous cattle. Navigators in the early 1800s were well aware of the iridescent slicks from seafloor seeps off the Santa Barbara Coast.

Most of the early production of oil was from surface seeps or by mining rather than from drilling wells. Apparently in 1855 or 1856, Andres Pico of the San Fernando Mission tried to refine some heavy seep oil from Pico Canyon near Newhall. Similar attempts were made at about the same time at Rancho La Brea, just west of a little town called Los Angeles. A further attempt was made at Carpinteria, near Santa Barbara, in 1857. Although not commercially successful, these early instances presaged the later development of California's important oil-producing districts.

Mining became the main method of extraction in the 1860s. At Asphalto in San Joaquin Valley, near what would later be the town of McKittrick, shafts and tunnels up to 300 feet deep were sunk into ledges, yielding a tarry oil. This they could refine to produce a bitumen much purer than that from the Caribbean island of Trinidad, then the major producer of asphalt. This material was used for roads and sidewalks in San Francisco, and as grease to skid logs in timbering operations. One shaft yielded a column of solid tar ten feet high and six feet wide, which was shipped intact for display in San Francisco.

Starting in 1861 at Sulphur Mountain near Ventura, tunnels were drilled at a slightly uphill angle. The miners made a trough in a tunnel floor with wooden planks, allowing water and oil to flow out. Tongue-and-groove redwood boards were used as siding to seal off water zones. Tunneling continued for decades, even after conventional drilling took over. One tunnel, dug by Union Oil in 1890, was 1,940 feet long. In at least one case, a mirror was set up to reflect sunlight into the tunnel for alignment purposes. Up to fifty-four of these tunnels were eventually dug, and some were still producing into the twenty-first century. In fact, some tunnels experienced a significant increase in production as a result of the 1994 Northridge earthquake, which may have opened up fractures previously sealed by asphalt.

Above: Entrance to salt marsh tunnel.
COURTESY OF CALIFORNIA OIL MUSEUM, SANTA PAULA, CALIFORNIA.

Below: Tunnel at Sulphur Mountain with a mirror at the entrance to reflect sunlight in.
COURTESY OF STEVE MULQUEEN.

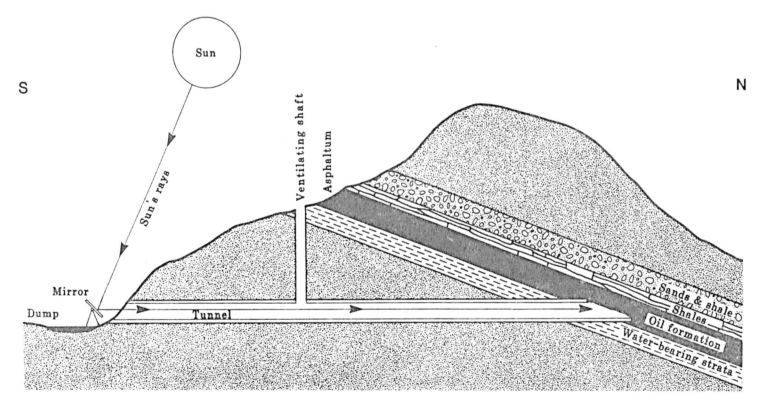

Above: Oil seepage on road cut near Santa Paula. The oil is very viscous and slow flowing because the lighter components have been lost through evaporation.
PHOTO BY AUTHOR.

Below: Portion of Ventura County showing Spanish grant ranchos investigated by Professor Silliman, including Rancho Ojai. Sulphur Mountain, where tunnels were dug for oil, is located in Rancho Ojai.
COURTESY OF STEVE MULQUEEN, MODIFIED.

Right after the Drake Well was drilled in Titusville, Pennsylvania, in 1859, the first attempts were made to produce oil commercially by drilling in California. The first well, drilled in Humboldt County in 1861, was dry. The first producing well was drilled near the town of Petrolia in the same county in 1865. The initial shipment of about 100 gallons of refined, light oil is said to have sold in San Francisco for $1.40 a gallon. Small amounts of up to fifteen barrels were shipped intermittently for the next two years.

THE EXPERTS ARE STUMPED

Such was the state of the petroleum industry in California when, in April 1864, a professor of chemistry and geology from Yale named Benjamin Silliman, Jr., arrived from the East. He came at the behest of another easterner, financier Thomas A. Scott, to investigate oil possibilities in California, Nevada, and Arizona. Having written a report that had encouraged investors in the Drake Well in Pennsylvania, Professor Silliman was highly reputed as a consultant to the oil industry. Silliman wrote reports on several Spanish grant estates, including the 18,000 acre Rancho Ojai near Ventura. He spoke glowingly and with some hyperbole about seeps, calling them "natural wells" and "rivers of oil." Large pools of underground oil had been found by drilling near seeps in Pennsylvania, and Silliman thought the same would be true in California. The Ojai report formed part of a prospectus for Scott's California Petroleum Company.

Scott's prospectus was prescient in that it claimed, in 1864, that oil would replace coal as the source of illuminant. It also correctly predicted that the market for California oil would include the whole Pacific region as far as Australia. Silliman recognized, although he did not name, the Monterey Formation as the source rock, an important step in understanding the geological habitat of oil in California. He said that the prospects for Ojai were better than those for Titusville before the Drake Well was drilled. Unfortunately, this highly optimistic prediction was based on rather cursory observations of seeps and surface geology, sometimes made from a buggy or the saddle of a horse.

A small boom, with several companies being created and shallow wells being drilled, occurred largely on the strength of Silliman's rosy projections. However, none of the wells produced commercially. The end of the Civil War in 1865 and increased production in Pennsylvania that drove prices down soon arrested the little boom. Silliman was accused of misleading investors with an overoptimistic report. There were even charges that he falsified chemical analyses of the oil to make it appear more valuable. Silliman was forced to resign his chemistry professorship at Yale. Eventually all of his predictions would be more than vindicated, but that would be decades in the future.

California's complex geology and the nature of its oil presented problems that Silliman and the other easterners could not have anticipated. Unlike the more or less "layer cake" geology of Pennsylvania, oil-producing beds in California are folded and faulted into contorted shapes. Oil might travel up a dipping bed or fault to a seep. A well drilled next to such a seep would miss the deep pool by a wide margin. Even if they did find oil, the early explorers were often defeated by the chemistry of the California crude, which made it almost impossible to refine by the methods of the day. The main product of refining in those days was kerosene, which was used for illumination. The heavy oil yielded a much lower percentage of kerosene than the light Pennsylvania oil, and the kerosene it did produce was of low quality for illumination purposes. The smelly, smoke-producing stuff ignited at an unsafe low temperature. It would sell only when the pure, eastern variety was in short supply and expensive.

NEWHALL: FROM GOLD TO BLACK GOLD

In 1842, a certain Francisco Lopez was supposedly napping under an oak tree in Placerita Canyon, near what would become Newhall in northern Los Angeles County. He dreamt that he was floating on a pool of gold, then woke up and proceeded to unearth some wild onions on whose roots he found little flakes of gold. This aptly named canyon thus became the scene of California's first "gold rush," six years before the big strike at Sutter's Mill. This little rush was soon forgotten, and Placerita Canyon went on to become a scene of Hollywood westerns. But there was something else in the ground, still unknown and untapped, far more fabulous than gold or movies. Lopez is said to have played a role in that discovery too.

The Newhall area with its many canyons was at the eastern edge of what was to become a swath of oil country stretching to the rich Ventura fields on the coast. Although part of the region originally touted by Professor Silliman in 1864, Newhall was not targeted by Scott because it was public land and could not be purchased like the ranchos to the west. This left Andres Pico and other locals, including Lopez, to pursue the seeps

Oil producing region stretching from Newhall to Ventura and beyond. Present day oil fields (green) and gas fields (red) are shown. Pico Canyon was discovered in 1876, Adams Canyon in 1888. The first offshore well in the U.S. was drilled at Summerland in 1897. Grey lines are the present system of highways and freeways.

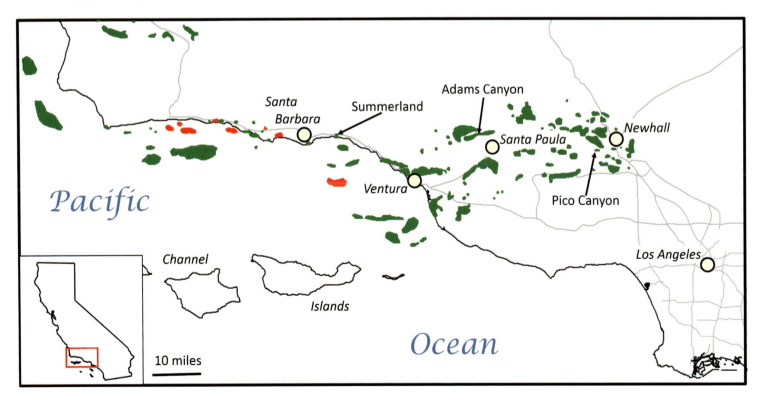

Top and inset: CSO Hill, named after California Star Oil Works Co., as it appears today (top), and in 1893 (inset). The Pico No. 4 discovery well is located off-frame, about 500 feet to the left of where this picture was taken. Virtually nothing remains of the derricks, bridge, shacks and shop buildings. The almost vertically dipping beds of sandstone, and the ruggedness of the terrain, were big problems for these early, intrepid drillers.
INSET PHOTO COURTESY OF CHEVRON USA.
MODERN PHOTO BY AUTHOR.

they knew about in the canyons. Pico, the son of the last Mexican governor of California, Pio Pico, teamed up with Edward F. Beale. Pico had been leader of the Mexican army at the Battle of San Pasqual, and Beale was the leader of the Americans. After Mexico lost the war Pico decided to stay on in California, and the two old soldiers became friends. They and several others including Colonel R. F. Baker harvested oil from pits and shallow hand-dug wells.

Soon, more explorers arrived with their primitive drilling rigs. In 1868 Francisco Lopez apparently showed H. C. Wiley, Sanford Lyon, and W. W. Jenkins a seep in Pico Canyon. This group drilled a well using a "spring pole" rig, which consisted of a fulcrum holding a flexible tree trunk with stirrups at one end in which men placed their feet to push the bit up and down. An emerald green oil was found. A few other wells were drilled, but at this time most of the production was still from pits and surface seepages. Canyon Country today is littered with place names from these early oil pioneers: Wiley Canyon, Lyon's Station, and others.

Dr. Vincent Gelcich, the coroner in Los Angeles and an in-law of Pico, was part owner of a mining claim in one of the canyons and wanted to build a refinery for the Newhall oil. In 1872 he went to San Francisco to promote oil over coal as a source for illuminating gas. A company called Metropolitan Gas works was dedicated to this idea and had built a refinery on Channel Street. Unfortunately, Metropolitan had two larger competitors that used coal, and price cutting in 1873 doomed the lightly capitalized company.

Gelcich saw greener pastures back in his home town, Los Angeles, where he touted the idea of a refinery. He got the editor of the *Los Angeles Star* to publish a telegram about a new oil strike at Newhall, alleging that a refinery venture could realize a profit of 70 percent. Soon, the Los Angeles Petroleum Refining Company was set up with local banker F. P. F. Temple as president, and Gelcich, Beale, Baker, and Pio Pico as major stockholders. By the end of 1873, construction of a refinery was underway at Lyon's station (or Petroliopolis), near Newhall.

In April 1874, the refinery had been built and was processing oil hauled in on wagons from Pico Canyon, eight miles away. Initial newspaper reports optimistically said that the kerosene was purer than that from

Above: Shop buildings, sheds, and shacks
that the workers lived in were crowded in
the narrow Pico Canyon with derricks
and pumps.

COURTESY OF CHEVRON USA.

Left: Inside the blacksmith shop at
Pico Canyon. Damaged or worn drilling
tools had to be repaired on site. Bits
and other tools were often improvised or
invented from scratch to meet the peculiar
needs of a well.

COURTESY OF CHEVRON USA.

Above: Demetrius G. Scofield, schoolmate of Mentry in Pennsylvania, was one of the founders of CSOW, and later was chairman of Standard Oil of California.

PHOTO COURTESY OF CHEVRON USA.

Right: The Newhall refinery as it appeared in 1880, a few years after it was moved to a new site next to the Southern Pacific Railroad. The year 1880 was also the year a new refinery was built on San Francisco Bay, so this picture shows the Newhall facility at the height of its operation.

COURTESY OF CHEVRON USA.

Pennsylvania. Soon, however, these pronouncements were replaced with silence. As in previous attempts, the kerosene was of poor quality, and the refinery shut down less than a year later. Several more attempts were made, some employing dubious "black box" methods. Companies came and went. Finally, a refiner from Pennsylvania named Joseph A. Scott arrived in January 1876 with his own secret process. His trip was paid for by Reuben Denton, who became the first San Franciscan to make a substantial investment in Newhall oil. Scott managed to produce kerosene that was better than any produced previously in California. Even though it still was not of the quality of the Pennsylvania product, its cheap price made it marketable.

The arrival of Denton and eventually other large investors on the scene changed everything. Previously, it was hard to get investors interested. Most of the investors before 1876 were from Los Angeles, a little town of 8,000 people, and there was not enough money to really make a go of it. Not only was the crude difficult to refine, the area was quite remote. There was no railroad, and the only way to get a vehicle to Los Angeles was through Beale's narrow, one-way toll road with 29 percent grades cut ninety feet into a mountain pass. More investment was needed. The obvious source was San Francisco (population 200,000), the prime market for illuminants.

Finally, Denton and Pennsylvania oil man Demetrius G. Scofield got San Francisco investors together to form the California Star Oil Works company (CSOW). Charles Mentry, a driller from Pennsylvania who was already active in the area, was hired by Scofield. Having drilled three wells with a spring pole that found small amounts of oil, Mentry now brought in a steam engine to power his rig. That made all the difference.

On September 26, 1876, Mentry's famous Pico No. 4 hit pay dirt at 370 feet, producing 25 barrels a day. This was the first commercial well in California. Amazingly, it was also the longest continually producing commercial well in the world in 1989 when it was shut in after 113 years of production. It opened up the Newhall area, which was then the state's most prolific producer. And it spawned a little community, now a well-preserved ghost town, Mentryville, named after the resourceful and determined driller.

Slowly, problems were overcome one by one. The original refinery was enlarged and moved a few miles to be next to the Southern Pacific Railroad, whose newly completed Bakersfield-Los Angeles route passed nearby in 1876. The kerosene was still of rather low quality, and in 1878 J. A. Scott was let go. A vexing problem that still remained was controversy over the titles to the land in Pico Canyon. This was caused largely by the use of mining laws to file claims, which did not work well for oil properties. From 1878 on the future of CSOW was clouded by a series of lawsuits between it and the duo of Beale and Baker. Until the dispute was settled the value of the company was in doubt, and its ability to attract investors and maintain its activities was inhibited. The stalemate continued until much bigger money appeared on the scene, represented by Charles N. Felton.

Arriving from New York in 1849 at the age of 17, Felton would in time become one of the outstanding financial and political leaders of nineteenth century California. Attracted to the goldfields like so many others he made his fortune not by mining but by being a merchant and banker in gold country towns like Marysville and Nevada City. He also served as an undersheriff and tax collector in Yuba County, and was later elected to the State Assembly. Moving to San Francisco in 1863, Felton soon joined the circle of influential financiers of Nob Hill. By 1868 he was the Treasurer of the San Francisco Mint. Later in life he would become a Congressman and then a U.S. Senator. It seems that Felton became interested in oil around 1877 by events at Moody Gulch near San Jose. In a few years his attention was directed south to the Newhall area.

Felton represented financial interests far more weighty than those of Denton, Scofield and the others. In 1879 he founded Pacific Coast Oil Company (PCO), probably for the express intention of taking over CSOW. The presence of this large moneyed interest may have prompted Beale and Baker to make a deal. PCO quickly took over CSOW and made it a subsidiary. This combination would

Above and inset: Newhall refinery as it appears today. Two of the original stills remain. The plaque (inset) on one of the stills says that the site was restored by Standard in 1930.

PHOTOGRAPHS BY AUTHOR.

Below: Charles Felton, the financier who founded Pacific Coast Oil Company (PCO).

COURTESY OF CHEVRON USA.

MENTRYVILLE

Driving west just a few miles on Lyons Avenue from the Golden State Freeway is a trip from the modern bedroom community of Santa Clarita to a nearly forgotten past. This was a past when steam was the motive power, travel was slow and difficult, and one had to find or make the necessities of life very close at hand. In the 1870s the oil field workers lived in cabins or even tents among the derricks. Pico Canyon was narrow, and in addition to the wells was crowded with the field office, a machine shop, storage sheds, and the workers' shacks. The nearest "big city," Newhall (population about 50), was seven miles away, which would have been quite a hike to get to work every day. Eventually the buildings got in the way of oil field operations, and a little company town was set up about a mile and a half down the canyon. The town was originally called "Pico," but soon acquired the name Mentryville. There was a boarding house for the single men, and little redwood cabins scattered about for men who had families. There was a bakery, and a stage came twice a day. The stage owner was probably happy about the town rule against liquor as it undoubtedly gave him a lot of passengers wanting to go to the "Derrick" saloon in Newhall. Recreation in Mentryville consisted of dances, picnics, tennis, and box-lunch socials.

Alex Mentry built a fine Pennsylvania-style mansion, with thirteen rooms, gas chandeliers, gas fireplaces, pull-chain toilets, alabaster wash basins, and no electricity until 1948. This house was occupied by Mentry and his successors off and on until 1994. With wives came children, and that meant a school had to be built. The school and the mansion, along with barns and a few other buildings, still stand.

After the oil field was abandoned in 1989 PCO's successor Chevron donated the town and surrounding land to the Santa Monica Mountains Conservancy. Now a well-preserved ghost town, Mentryville is used for movies, weddings and other events. Up the canyon is a replica of an old wooden derrick, some discarded equipment, and a monument at the location of the famous Pico No. 4 well, which is on the National Historic Register. Mentryville in a sense lives on, serving as a reminder of the pioneer days of petroleum.

Opposite: Mentry's mansion.
PHOTO BY THE AUTHOR.

Above: Mentryville from the air. A good
portion of the population seems to be posing
for the picture.
COURTESY OF CHEVRON USA.

Left: Mentryville today.
PHOTO BY THE AUTHOR.

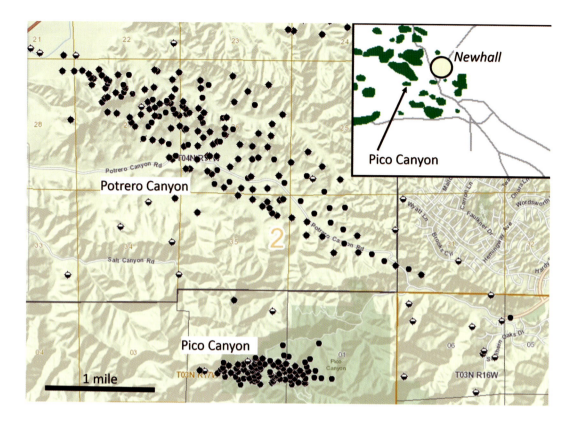

Right: Nineteenth and twentieth century oil fields: The many, shallow (mostly less than 1,600 feet) wells are closely spaced in the narrow, rugged Pico Canyon. By comparison, the Newhall-Potrero Field just to the north, discovered in 1937, has much deeper producing zones (6,400 to 14,500 feet) and more widely spaced wells.

MAP MODIFIED FROM THE CALIFORNIA DEPARTMENT OF OIL, GAS, AND GEOTHERMAL RESOURCES (DOGGR).

Below: The first field discovered in the Los Angeles Basin was at Puente Hills in 1880. It was later combined with other discoveries as one field, Brea-Olinda. The next two fields discovered in the Los Angeles Basin were Los Angeles City (1892) and Whittier (1896). This was only the beginning. Gray lines on this and subsequent maps represent the present-day freeway system for geographic reference.

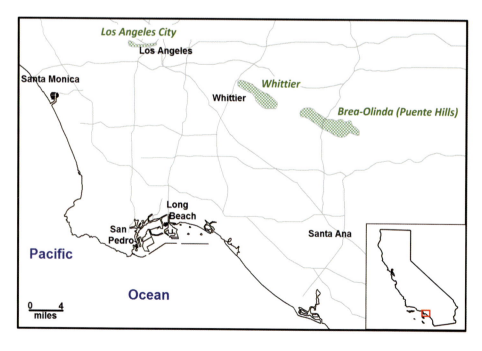

later form an integral part of what would eventually be called Chevron. The first pipeline in California was completed in 1879, a 7 mile, 2 inch affair to carry oil to the refinery. By the 1880s Newhall production was up to 100,000 barrels a year.

In 1880 a new refinery was opened by PCO at Alameda on San Francisco Bay to handle oil from newly completed wells at Moody Gulch. Much of the Newhall oil also went to Alameda. In 1890 the Newhall refinery was closed. By 1900, when Charles Mentry died, 70 wells had been drilled. Another 28 wells were drilled after 1900, the last one in 1969. Production continued in several other canyons, among them De Witt, Townsley, Wiley, and Rice. Production continues today in Potrero Canyon, just two miles from the original Pico Canyon Field, as well as at the site of the original gold rush, Placerita Canyon. Pico Canyon produced about 3 million barrels in its lifetime, and as of 2009 all of the canyons around Newhall had produced over 150 million.

BOOM AND BUST

The 1876 discovery at Newhall led to the first real California oil boom about four years later. Annual production of 12,000 to 15,000 barrels in 1876-1878 rose to almost 100,000 by 1881. Until about 1885 most of the oil came from Newhall. The 1880s saw discoveries in rapid succession: Puente Hills southeast of Los Angeles, Summerland in Santa Barbara County, and Adams Canyon near Santa Paula. Then there were the first glimmers of what was to come in San Joaquin Valley. Production soared to nearly 700,000 barrels in 1888.

Oil was discovered in 1880 in the Puente Hills by drilling 100 foot wells near seeps along the crest of a tightly folded anticline, much like that of Pico Canyon. This was the next oil field discovered in Los Angeles County after Newhall. These early shallow wells yielded heavy crude and production was small. In a few years deeper wells were being drilled. Very light oil (API gravity 30 to 35) was produced, showing that light oil could be found in California by drilling deep. By 1900 about 60 wells had been drilled to an average depth of about 1,200 feet. Until 1894 most of the oil was shipped to Los Angeles. After that it was shipped to Chino to be refined, the residuum being supplied as a fuel to a nearby sugar beet factory. This is an example of the supplanting of coal by oil as a fuel for steam engines. Other discoveries were made at Olinda in 1897 and Brea in 1899. Ultimately these and other fields merged into the Brea-Olinda Field as all of the intervening areas were proven productive.

Oil seeps along the coast southeast of Santa Barbara had been long known by the Chumash people. Prospectors dug for oil, finding a heavy (API gravity 7) oil that had little commercial value. The first well, drilled in 1886, found sub-commercial amounts of lighter oil. A small strike was made near the little coastal town of Summerland, and by 1894 derricks dotted the beaches and bluffs nearby. Correctly surmising that the field extended offshore, drillers began building piers for their derricks in 1897; these were the first offshore oil wells in the United States. The Southern Pacific rail line from Los Angeles to Santa Barbara ran right through the

Summerland Oil Field. The railroad decided to fuel their locomotives with the field's heavy oil rather than with coal. One pier was built 1,230 feet into the ocean with 19 wells along it to produce oil for the trains.

The first real oil wells in the San Joaquin Valley were drilled in 1878 and 1879 near the future town of McKittrick, where asphalt had been mined for more than ten years. These early wells were not very successful. Finally in 1887 the Wild Goose well came in at 10 barrels a day and the oil field town of "Oil City" existed for a few years. Then in 1889 the discovery well was drilled in the Sunset Field. Rail lines were extended to McKittrick. All of these developments foretold a bright future for oil in San Joaquin Valley, but by 1889 and 1890 the oil industry in California was depressed. The real breakthroughs in the valley came later when the next boom was already going strong: the Shamrock Gusher at McKittrick (1896), the Blue Goose Gusher at Coalinga (1898), and the Kern River discovery (1899).

Perhaps the most important strike in the 1880s was near Santa Paula in Ventura County. This discovery led to the founding of one of California's greatest oil companies, Union Oil, produced the first gusher in California, and finally gave truth to Silliman's twenty-year-old vision of "rivers of oil." It is also a story of one man, Lyman Stewart, who risked everything, made and lost many fortunes, overcame sickness and defeated determined opposition. Stewart was the one whose dogged persistence carried the day for Union Oil. Stewart also had the vision to help prod California's industries to convert from coal to oil, and he

Top, left: Central Oil Company had 2,700 acres in the Puente Hills near Whittier, and by 1900 had 800 wells. It was one of the more prominent companies in the field. This photograph was taken in 1895.
COURTESY OF THE CALIFORNIA STATE LIBRARY.

Top, right: Wooden derricks on piers, Summerland. This is where the first offshore well was drilled in 1897. By the time this photo was taken many piers had been built, with multiple derricks on each pier. By 1939 all of this had been washed away by storms.
COURTESY OF CALIFORNIA OIL MUSEUM, SANTA PAULA, CALIFORNIA.

Below: Lyman Stewart.
COURTESY OF CALIFORNIA OIL MUSEUM, SANTA PAULA, CALIFORNIA.

Above: Oil field in a canyon near Santa Paula, looking down a steep tramway that was used to bring workers in and out each day. This illustrates the rugged topography that the early oil developers had to negotiate.
COURTESY OF CALIFORNIA OIL MUSEUM, SANTA PAULA, CALIFORNIA.

Opposite, top: Oil wells on small ledges atop cliffs.
COURTESY OF CALIFORNIA OIL MUSEUM, SANTA PAULA, CALIFORNIA.

Opposite, bottom: Santa Paula in 1889. This is right after the Adams No. 16 gusher of 1888 and before the founding of Union Oil in 1890.
COURTESY OF CALIFORNIA OIL MUSEUM, SANTA PAULA, CALIFORNIA.

formed a company that innovated oil field tools and machinery for decades.

Lyman Stewart, a man of contradictions, was born in 1840 of a deeply religious, Scottish family that was not wealthy, yet highly respected. Lyman's father was one of two tanners in the rural area of western Pennsylvania that was to become the scene of a fantastic oil strike some 19 years later. Following the custom of the times, Lyman was apprenticed at age 12 to follow in his father's footsteps as a tanner. By 1859, when the famous Drake well was drilled in nearby Titusville, 19-year-old Lyman Stewart was establishing himself as a full-fledged tanner, but he loathed the job. He hated the idea of spending his life as a tanner, and wanted to come up with money to allow him to do what he really wanted: become a missionary. Oil seemed the answer. He had $125, a substantial sum for a 19-year-old at that time, and with partners invested all of it in an oil lease. The trouble was that the partners did not keep back any money to drill a well, and they ended up losing the lease. Two years later he had saved enough to invest in another lease and actually drilled a well. This time the

price of oil nosedived, and Stewart again lost his investment.

Stewart then spent the Civil War in the Pennsylvania Cavalry. In 1865 he came back to find his hometown turned into an oil boom madhouse. He took a business course and opened up an oil leasing office. Of great advantage was his intimate knowledge of the local terrain and farmers' properties. This priceless asset he had gained from his days as a tanner riding all over the county on horseback to pick up hides and deliver leather. He learned to buy small 1/64 interests in oil leases so that he could spread his capital and minimize risk. By 1872 Stewart and his brother Milton had amassed a fortune of $300,000, a princely sum in those days. Unfortunately Stewart got involved in a scheme to sell farm implements, and he lost every penny as well as his house. He was forced to take a low-paying job just to support his family. This might have been the end of the story as far as the history books are concerned, but Lyman Stewart was not a quitter.

It seemed as though providential events occurred in Stewart's life when least expected, and when they were most needed. During one

of his plush periods he had befriended two young brothers from Maine, James and Harvey Hardison, who were oil field laborers good at fishing for lost drilling tools. Stewart helped them start a business, and although he did not profit monetarily from this, he made a connection that changed his life. Just when things seemed darkest, a third Hardison

brother showed up in Pennsylvania. Wallace Hardison, who had made a fortune in the West from railroad ties, came to meet Stewart based on kind words about him from his two brothers. Wallace proposed a partnership with Stewart: Stewart knew about the oil business, said Hardison, and he (Hardison) had the money. Stewart's confidence was restored, and the partners went on to make yet another fortune from oil.

One of the Hardison-Stewart partnership's projects was part of the new field in Bradford, Pennsylvania, which eventually produced 100,000 barrels a day from 7,000 wells. This kind of production, 80 percent of the U.S. total in 1881, caused the price of oil to plummet. Big buyers like John D. Rockefeller's Standard Oil monopolized pipelines, railroads, and other distribution facilities. The wellhead oil price was forced down to a ruinous 8 cents a barrel. In 1882 Stewart was approached by I. E. Blake, an acquaintance who had gone to California and gotten involved in Pico Canyon. Blake suggested that Stewart and Hardison pull up stakes and look for oil out west. Stewart sold out most of his Pennsylvania interests, netted $70,000, and made the trip in 1883. Although Hardison initially decided to go to Kansas and try his hand at farming, he was later persuaded to rejoin Stewart in California.

Why would Stewart, at age 43, make such a decision when the total production of all California fields since the industry started was about equal to what the Bradford Field put out in five days? Why would a conservative, cautious businessman abandon a sure source of income for a risky adventure in an unknown place? Probably Blake did not tell him about the complex California geology, or the difficulty in refining the crude. Stewart was to find out these things, and more, for himself. Stewart's move was probably influenced by the monopolistic tactics of Rockefeller and others in Pennsylvania. It was a bold decision, regardless of the motivation. It is another example of Stewart's risk taking, and it caused heartache and seemingly endless struggles. But it paid off spectacularly.

In California, Blake showed Stewart the Pico Canyon Field. Stewart saw the success of some of his fellow Pennsylvanians, including Mentry, Joseph Scott, and former schoolmate Demetrius Scofield. Blake proposed that Stewart and Hardison sublease unproven land of the Pacific Coast Oil Company and drill there. Stewart and Hardison ended up drilling six dry holes in Pico Canyon and two others near Santa Paula. Desperately low on funds, Stewart asked Blake if they could drill on proven land owned by PCO. Back in Pico Canyon, they spent the last of their money to finally bring in a producer, Star No. 1, at 75 barrels a day. Having no money left to develop the field, they had no choice other than to take a cash offer from Blake for the well. This taught them another lesson: drill your own leases so that you can make your own decisions. Stewart and Hardison decided to use the money to make down payments on mineral leases near Santa Paula, and then used the leases as collateral to borrow money to drill wells.

In those years oil was being produced around Santa Paula mostly from tunnels. The first successful well, Old Adams in Adams Canyon, had been drilled in 1875 with a small production of 2 or 3 barrels a day. In 1884 Stewart and Hardison drilled several wells and got a small amount of production. Unfortunately their first two wells were being drained by later ones. The oil was very viscous and was useful only for asphalt. Then, in 1884, they heard about the light oil from deep drilling in the Puente Field. Another Pennsylvania associate, W. E. Youle, had drilled a 1,600 foot well there and was visited by Stewart and Hardison, who wondered how this oil could be so much lighter than that from nearby shallow wells. They lacked modern geological knowledge, yet they could still test the idea of drilling deep in Adams Canyon. Returning there they drilled several more wells with partial success.

Finally, in 1888, the first gusher in California, Adams No. 16, came in at 500 barrels a day with oil spouting over the top of the derrick. Four years later this modest gusher was topped by the Adams No. 28 at 1,500 barrels a day. This one took the drillers by surprise, and they were unable to contain the oil. The resulting river flowed all the way

to the ocean, in a sense validating Silliman's rather colorful statements of twenty-four years before. This success helped establish the Stewart and Hardison partnership as a force in the state's petroleum industry, and put Santa Paula on the map as an important oil town.

As important as the discoveries at Puente, Summerland, Adams Canyon, and the other places were for the future, the boom was short-lived, production dropping more than 50 percent in 1889. Nonetheless California had gotten its first real taste of the oil bug. California petroleum was becoming more valuable as its main use shifted from illumination to fuel. New technologies were being developed, such as better refining methods, nozzles to allow oil to efficiently burn in boilers, pipelines to move heavy oil, and tanker-ships. Oil trade with Pacific nations such as Mexico and the Kingdom of Hawaii was opening up. The debacles of the 1860s were becoming dim memories. The stage was set for the next boom, ushered in by a remarkable discovery in the city of Los Angeles, which would dwarf everything that had come before. It would begin a long-term upward trend of production in California (with short-term fluctuations) that would last from 1892 until about 1982, when it topped out at over 420,000,000 barrels a year.

Opposite page: Union Oil refinery in Santa Paula. This refinery began operations in 1887 under one of Union's predecessor companies. It remained Union's only refinery until the more advanced one at Oleum (San Francisco Bay) was opened in 1896. The Santa Paula refinery was destroyed by fire just four months later.
COURTESY OF CALIFORNIA OIL MUSEUM, SANTA PAULA, CALIFORNIA.

Below: Annual oil production in California, 1875-1900. Dates of major discoveries are shown along the bottom. The second boom in the 1880s and beginning of the third boom in the 1890s are shown.
DATA FROM REDPATH, 1900.

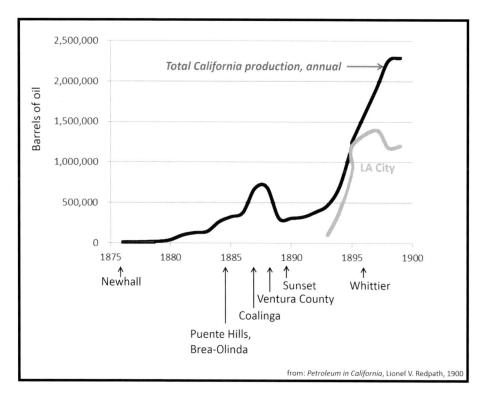

from: *Petroleum in California*, Lionel V. Redpath, 1900

THE NEXT BOOM:
LOS ANGELES CITY
AND BEYOND

In 1857 a small hand-dug well at Third and Coronado Streets on the west side of Los Angeles began providing heavy oil and asphaltum to oil the streets in the town. Little or no other successful drilling is recorded until 1890, when prospectors Maltman and Ruhland achieved production of a few barrels a day. Then, in 1892, the same year as the Adams No. 28 gusher, a down-on-his-luck Colorado prospector named Edward L. Doheny and his partner Charles A. Canfield began digging a shaft on a lot at the corner of Patton and Colton Streets. No trace remains to indicate the spot, now somewhere in the parking lot of the municipal Echo Park Swimming Pool. This four by six foot shaft was dug with picks and shovels. They encountered oil seeps as they dug, but when they reached 155 feet they had to stop because of toxic gas in the hole. Then Doheny used a eucalyptus trunk with a sharpened end as a percussion drill to deepen the well another 70 feet. They hit the oil reservoir, and began producing seven barrels a day. Thus began the Los Angeles City Field, which caused California production to skyrocket from under 400,000 barrels annually in 1893 to almost 2 million by 1897.

A real estate promotion had occurred here a few years before, and the whole area had been subdivided into small lots, some of them only 50 by 150 feet. Anyone who owned a lot and $1,500 could drill a well. Even without the money it could be done by attracting any of a hoard of investors. Of

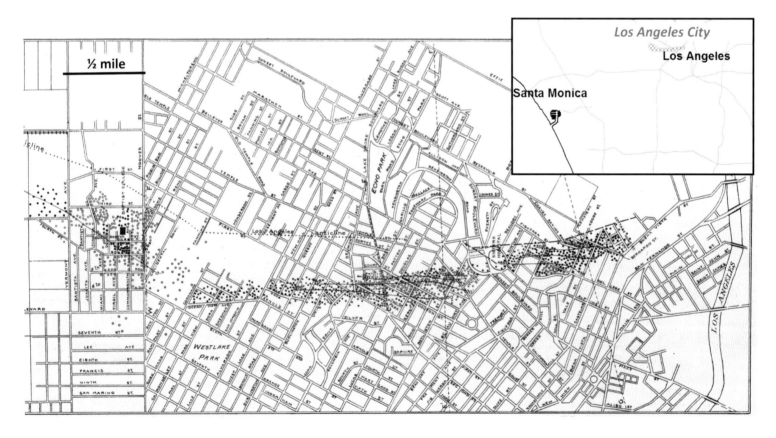

course one would have to drill before the neighbors drained the reservoir. If a neighbor drilled two wells, a person would drill two, or even more, on a tiny lot, just to keep up. In three years about 300 wells were sunk, mostly on a trend to the northeast. When they reached Victor Street, it appeared as if a fault terminated the productive reservoir, so the prospectors went back to the original discovery site and began moving west. In 1896 another discovery was made east of Victor Street, near Sisters' Hospital (near what is now Elysian Park).

At its peak the field was about three miles long east to west, stretching from Elysian Park on the east, passing north of Downtown and south of what is now Dodger Stadium, extending to about Vermont Avenue on the west. The field was only about 1,000 feet wide. Peak production of 1.8 million barrels from about 1,150 wells occurred in 1901. After that production declined rapidly. Only two new wells were drilled after 1915. By 1961, 93 producing wells remained and by 2015 only a few stripper wells were left.

Bimini Baths. The street along the left side of the spa is now called Bimini Place. The trolley tracks in the background may still be seen imbedded in the street's pavement and in a private driveway. Rayfield Apartments on the corner still exists.

In the first few years what had been a residential area became a forest of derricks. In some places one could walk from one derrick floor to another. There were no regulations at that time about well spacing or subsurface rights. At one point the city council restricted the drilling of oil wells in some areas. Suddenly people began drilling "water wells." Mishaps unavoidably occurred. In 1907 an oil tank made of redwood burst. Oil flowed down the street into Echo Lake, where it caught fire and burned for several days. Criminal behavior, such as stealing tools, or oil out of tanks, was common and difficult to prevent. Competition was intense among the many operators and sabotage was not unknown.

One well near Vermont Street never found oil. Instead, at 1,750 feet, seemingly inexhaustible amounts of hot water with dissolved sodium bicarbonate, prized for mineral bathing, began flowing. Thus was born the Bimini Bathhouse, which lasted from 1903 until the 1950s. Founded by dentist David W. Edwards, the spa was named after the tropical island that was supposed to have had the fountain of youth. Located on Bimini Place near Second Street, the enormous building had concrete tanks that could hold half a million gallons, and large gates that allowed water to quickly re-fill the pools every day. There were over 500 dressing rooms, separate floors for men and women, three large pools for athletic swimming and

diving, over 50 private baths, a café, and other amenities. Patrons could get Turkish baths, steam baths, and all sorts of "hydropathic" treatments. Dr. Edwards built a hotel across the street, and supplied it with the mineral water. Other establishments later sprang up around the spa, most notably the Palomar Ballroom, where the likes of Benny Goodman, Tommy Dorsey, and Glenn Miller entertained. They had marathon dances in the street in front of the spa. Bimini was quite a social spot and all you needed to go there was trolley fare and the $0.25 price of admission. Today the old hotel building soldiers on as a treatment center for alcoholics. The spa is long gone. The only memories are a street name and old trolley rails imbedded in the street's pavement.

The boom spawned hundreds of oil companies. The Los Angeles Oil Exchange was founded in 1899 to provide a place for stocks to be bought and sold. Dozens of advertisements could be found in newspapers and magazines. At first only stocks of companies with producing wells were allowed to be traded. However, the public's demand for oil stocks, any oil stocks, was insatiable. The California Oil and Stock Exchange, founded in 1900, decided to include non-producing companies; later the Los Angeles Oil Exchange followed suit. The Los Angeles Oil Exchange merged with the San Francisco Stock and Bond Exchange in 1957 to form the Pacific Stock Exchange, which maintained trading floors in both cities until 2001.

Among the countless companies was the Women's Pacific Coast Oil Company, which was run exclusively by women except for a consulting engineer named H. Hawgood, who supervised drilling operations. While men could invest in the company they were not allowed any role in management. Like companies today, the Women's Pacific Coast Oil Company had operations in many oil fields, including the Los Angeles City Field, Placerita Canyon in Newhall, Summerland in Santa Barbara, Brea-Olinda, and the new and promising Kern River District. This comprises nearly all of the major oil-producing areas of California at the time.

By 1900, new ways of exploring for petroleum were beginning to displace the early method of simply looking for seeps, which in large part was the underlying method used by Silliman in the 1860s. Fields such as Los Angeles City and Pico Canyon showed the importance of anticlines and, especially in California, faults. The new ideas can be found in the following comment by Professor W. L. Watts of the California State Mining Bureau, published in 1900:

> …in a general way, it may be said that the oil lines, or lines along which remunerative wells may be found, follow the strike of the axes of folds in the rocks, or the course of faults which have isolated blocks of strata inclosing the oil-yielding rocks…the tracing of oil lines in this state and the development of oil fields, necessitates a competent knowledge of structural geology, without which the risks of oil mining would be greatly increased.

However, old ways die hard. It would take more than twenty years for this new approach to take hold. Eventually this use of the scientific method, as applied to geology, would lead to discoveries of oil fields with no surface expression, in places totally unexpected. As we shall see, the resulting yield exceeds by orders of magnitude that previously found by looking for surface seeps.

Production from the Los Angeles City Field topped 1,000,000 barrels in 1895 and following years, fueling the third California boom and far surpassing the previous boom of the early 1880s. In the late '90s production from the gushers at McKittrick and Coalinga drove the boom still higher. Then, in 1899, the Kern River Field was discovered, setting the stage for even greater glory in the twentieth century. Another factor was just on the horizon: the rise of the automobile and the airplane, which would increase demand for petroleum exponentially, and shift the major usage from lighting and manufacturing to transportation. At the same time the Los Angeles City Field and Kern River Field would demonstrate the need for better control of oil field practices to maximize production and field life, in order to meet future demand.

Below: Advertisements for oil stocks, about 1910. Such advertisements were common in California publications of the time.
FROM OUT WEST MAGAZINE.

Bottom: Stock advertisement for the Women's Pacific Coast Oil Company.
FROM OUT WEST MAGAZINE.

So, she learned how to consolidate the operations of smaller, less efficient operators. By 1901 she had 14 wells altogether producing 50,000 barrels a month, about 2.5 percent of the state's total production. She achieved this by personally managing the business and looking to every detail including purchase of pipe and hiring of workers. After a day's work she would do the bookkeeping and give piano lessons to earn money to invest in more wells. At first she sold her oil through brokers. Later she took over that part of the business too, selling directly to hotels, factories, the local light company, and commuter railroads. Summers had 40 horses, 10 wagons, and a blacksmith.

Emma Summers was highly intelligent and organized. She used the laws of supply and demand to dominate the market in a volatile environment in which prices varied between $1.80 and ten cents a barrel. In the process she bought out other, failed operators for bargain prices. She did all of this without taking in any other investors. It was with good reason that Sunset magazine said in 1911, "There are men in Los Angeles who do not like Emma A. Summers." Yet she was fair to her customers and employees, and became known as the "Oil Queen of California."

Banker's daughter Emma McCutchen wanted to make music her life, having graduated from the New England Conservatory of Music. She married carpenter Alpha Summers, and they moved from Kentucky to Los Angeles, where Mrs. Summers intended to give piano lessons. However, a completely different line of work was destined to make her famous, not to mention very rich. The new Summers home was just a few blocks from where Doheny and Canfield would later drill their discovery well. When the oil field came in the oil bug bit Emma Summers, and it bit hard. In 1893 Summers paid $700 for a half interest in a nearby well. Unfortunately the well had casing problems and lost tools, and she had to borrow another $1,800 to save the well. "Night after night, by the light of a flaring torch, she hovered over it, as if it were a sick babe's cradle." Finally the babe recovered and became a producer.

The bug bit harder. Summers invested in another well, then another. When she was $10,000 in debt (easily four or five years' wages in 1900), she thought that if she could get her money back and that much more, she would quit. She made the $20,000, yet she didn't quit. She used boilers to remove sediment from the viscous oil; such tasks were more efficiently carried out by an operator who could combine the production of several wells.

Eventually the Los Angeles City Field faded, so Summers diversified. She got into real estate, buying ranches in the San Fernando Valley, apartment houses, and a mansion on Wilshire Boulevard where Bullock's Wilshire now stands. She converted one building into a hotel and named it the Queen. Then she purchased several of the new-fangled movie houses.

Later in life Emma Summers lived in nice style at the Biltmore Hotel and the Alexandria Hotel. She finally passed away in Glendale in 1941 at the age of 83. Not much is known of her husband, who died in 1939, and less is known of any children. Los Angeles had become a great city by 1941, and Emma Summers had something to do with it. She has been gone for three quarters of a century now. Few of today's citizens of Los Angeles know anything about her, but for anyone who cares to know her life's story, she is a true representative of the American Dream, a person who started a business and thrived through her own resourcefulness and hard work.

Above: Emma A. Summers, the Oil Queen of California.

Right: Los Angeles City Field, showing derricks with legs almost interlacing in some cases, and tanks belonging to individual owners. Entrepreneurs such as Emma Summers bought out these small operators and consolidated them into larger, more efficient units.

COURTESY OF CALIFORNIA OIL MUSEUM, SANTA PAULA, CALIFORNIA.

TWO OIL GIANTS OF CALIFORNIA

Two large oil companies, quintessentially Californian, have had an enormous impact on the petroleum industry world-wide, not just in California. As we have seen, both Chevron (Standard of California) and Unocal (Union) had their origins in tiny, shaky companies of the 1870s and 1880s. One company, Chevron, is a "major" that was originally part of the great Standard Oil trust, while the other, Unocal, was always known as an independent, albeit a very large one. Both companies grew in the twentieth century in part by absorbing others, Standard taking Gulf and Texaco, and Union merging with Pure Oil, so that their operations expanded far beyond California.

In 1885, the Stewart-Hardison partnership that was to become Union was drilling deep wells (up to 3,000 feet) based on what they learned from the Puente Field. However they produced only 4,806 barrels that year. Costs of drilling and operating the fields exceeded revenue, forcing them to borrow to stay afloat. To reduce costs, Stewart decided to use oil that they produced to run their boilers, rather than coal. Through trial and error his crews developed a nozzle that made this practical. This was the beginning of the revolution in which oil replaced coal as an industrial fuel in California. To reduce cost of shipping oil to San Francisco, they built a four-inch pipeline to the Pacific Ocean, the first of its kind in California. Production increased to over 35,000 barrels in 1886, about 10 percent of the state's total. Even so, the partnership teetered on the edge of collapse: they needed more money.

This is where Thomas R. Bard came in. Bard had drilled wells in the unsuccessful Rancho Ojai venture in the 1860s, and later made a fortune developing and selling real estate. Bard came into the partnership as an "angel." A group of four companies was formed with Hardison and Stewart having majority interest, and Bard more or less in operational control. The four companies were merged in 1890 to form Union Oil Company. The stage was set for an immense struggle between the cautious Bard and the aggressive Stewart, who wanted to get as many leases as

possible. They were still in debt, and several near disasters were barely averted even after the gushers at Adams Canyon. Production increased to 121,000 barrels in 1887 and over 236,000 in 1888. Bard wanted to simply produce and sell crude oil and let others do the refining, distribution and marketing. Stewart had much grander plans.

Stewart advocated tirelessly for conversion to oil as fuel for industries in California. He met fierce opposition from steamboat inspectors, railroads, and from his partner Thomas Bard. There were many setbacks, including steamboat explosions supposedly caused by fuel oil and railroad locomotives that could not develop enough power on oil. Gradually the technological problems were overcome, and the use of oil in engines surpassed its use as an illuminant. Stewart also formed a company to supply drilling tools to the industry, a move vehemently opposed by Bard. This and many other disagreements fed a steadily increasing acrimony. Finally, in 1898-1899, matters came to a head. Bard precipitated a fierce stockholder fight for control. The Stewart interests, when all of the shares were counted, held 50.6 percent. At last Lyman Stewart was in undisputed control, and from that time was free to put his plans into motion to create one of the greatest independent petroleum empires of all time. Although Stewart did not become a missionary, he founded the well-known Bible Institute of Los Angeles (Biola University).

Above: Thomas Bard.
COURTESY OF CALIFORNIA OIL MUSEUM, SANTA PAULA, CALIFORNIA.

Below: Union Oil Headquarters in Santa Paula, California. The hardware store on the ground floor was owned by the company. Kerosene lamps that burned Union's products are prominently displayed in the window. The horseless carriage, which would become the industry's mainstay, was still in the future.
COURTESY OF CALIFORNIA OIL MUSEUM, SANTA PAULA, CALIFORNIA (WHICH NOW OCCUPIES THIS BUILDING).

*Pacific Coast Oil Company built this
refinery at Alameda Island in 1880.
It would be expanded many times by
Standard of California.*

COURTESY OF CHEVRON USA.

The Pacific Coast Oil Company (PCO),
nucleus of what would become Chevron, pursued a different course than the brash Lyman
Stewart's Union Oil. Where Union acquired as
many leases as possible, precipitating numerous financial crises, PCO relied on its wells
in Pico Canyon Field and nearby. Although
Union's market was mainly in Southern
California, they would sell wherever they
could. PCO sold in the San Francisco area,
using product from their Alameda refinery.
Although this refinery had been built largely
to process the Moody Gulch oil, it fortunately
was located near the Southern Pacific Railroad
rather than at the end of a small narrow
gauge railroad near the Moody wells. When
the thin Moody sands played out they were
still able to ship their Pico oil north to
Alameda. As a result the smaller refinery in
Newhall was eventually phased out of service.
Pico production was not enough to satisfy
the San Francisco market, so prices remained
high. PCO's main competition was higher
quality eastern oils, marketed primarily by
Rockefeller's Standard Oil of Iowa, which
was on its way to becoming the dominant
marketer on the west coast in the 1880s.

On the production side, PCO concentrated
mainly on in-field drilling and new technology
to improve success. For example, the first
time in California that a diamond bit rotary
rig was used for oil drilling was in Pico
Canyon in 1889. The bit, a hollow cylinder
edged with diamonds, cut a straight hole.
However progress was much slower than
with cable tools, and the bit had trouble
with cobbles, which were ever-present in
the Pico formations. A side benefit of this
bit, which was similar to those used in
hard rock mining operations, was the ability
to take a core, making it possible to log
in detail the characteristics of strata
encountered by the well. This gave Edward
North, PCO's manager of the field, an idea.
He told Mentry to make such logs for all
future wells, and reconstruct them as best as
possible from memory for old wells. This
enabled them to construct geologic cross
sections that could be used to predict
depths of oil-bearing sands and give the best
locations for new wells.

Until 1894 the combined production of
PCO and its rival Union accounted for up to
75 percent of California's total. At this point

the Los Angeles City Field exceeded the totals of both companies and caused serious oversupply problems, especially for Union's southern California market. The heavy Los Angeles oil, unfit for refining into kerosene, was used as fuel oil in the local market. Union first tried to make a deal in which it would market and distribute the oil of many of the hundreds of small producers. The locals refused, causing Union to look for markets other than Los Angeles for its products. PCO in turn realized that its San Francisco market could be jeopardized as well if Union tried to shift its sales there. There was also a brief threat of substantial imports of newly-discovered oil from Peru.

To solve this marketing problem PCO approached Union with the idea of building a tanker ship to carry oil from both companies to San Francisco. PCO's reasoning was that since it could not supply all of the needs of San Francisco with its Pico oil, allowing Union to supply the rest would discourage competitors like Standard of Iowa. It would also give PCO the benefit of economy of scale. Union accepted the bargain, and in

1895 PCO built the *George Loomis*, the first ship in which the hull itself was the tank, divided into six sections by partitions, that carried the oil. Earlier tankers had tanks fitted into their holds, wasting space and adding weight. Thus PCO began a beneficial, although rather testy relationship with the rival it had helped to create a dozen years before. This lasted until the end of the century, when the existence of PCO as an independent company came to an end.

Standard Oil Company of New Jersey and the Rockefeller trust absorbed PCO in 1900-1901. PCO and Standard of Iowa, the Trust's marketing arm in the West, were merged to form Standard Oil Company of California in 1906. The head office in New York was firmly in control, and all significant decisions had to be approved by it. Standard of California became independent when the Standard trust was broken up by the U. S. Supreme Court in 1911. The merger with Standard of Iowa had provided the company with new assets and a strong marketing capability. Standard was free in 1911 to move forward in its own right as a major player in California's oil industry.

The George Loomis (on the left), was built by Pacific Coast Oil Company in 1895.
COURTESY OF CHEVRON USA.

Wooden pumpjack near Taft, photographed in about 1977. A relic out of its time, the pumpjack is shown still rigged up to a well.
COURTESY OF CARL F. BAGGE.

In the early days of wide-open wildcatting, poor-to-nonexistent record keeping, and many small operators, wells were often completed haphazardly. One operator could easily ruin the adjacent wells of another. To make matters worse, abandoned wells were often not properly plugged, so that any sands they penetrated could potentially be flooded with water. In 1904 production declined from the previous high of 17 million barrels to 14 million barrels, and many wells along the southern edge of the field had to be abandoned. In 1905, Colonel John Carter, making a reconnaissance of the field for Standard Oil, said:

> No matter what the production...was or what the hopes of the producer are, death and destruction surround that field, and it will only be a year or two at most, when it will be numbered with last year's snows and be forgotten.

Carter conveyed a similar pessimistic appraisal of other fields in California.

What could they do? Various attempts were made to seal off the space between the casing shoe and the wall of the hole using chopped rope, clay, brick chips, and cement. However, no matter what an operator did, he was often at the mercy of what his neighbor did. Disputes led to lawsuits and perhaps other less "civilized" actions. Gradually operators came to the realization that something general had to be done or most of the petroleum resources of California would be forever lost. The first attempt at legislation in 1903 failed because it provided no mechanism by which the law could be enforced. Additional laws passed in 1909 and 1911 had the same problem.

Finally some operators themselves got together in 1912 and formed the Kern County Oil Protective Association for the Midway-Sunset Field. It was largely a repository to which operators were supposed to furnish drilling records. However, some operators, including Standard Oil, did not join this organization. In 1914 the Coalinga Water Arbitration Association was formed with a certain amount of enforcement power over its members. These two organizations were of course voluntary and local. There was still no consistent way of attacking the water infiltration problem for all of the many oil fields in California.

The State Mining Bureau, formed in 1880, was mainly a clearing house for information about mineral resources in the state. In 1914 the State Mineralogist, Fletcher Hamilton, decided to make a survey of the petroleum industry. For this job he appointed Roy McLaughlin, an engineer who had worked at the mining camps of Bodie, California, and Manhattan, Nevada, and had been a geologist in the Taft oil region. McLaughlin's survey found that water infiltration had damaged many oil fields in California.

As a result of this survey Hamilton proposed a new statewide agency to ensure that all operators did everything possible to prevent water infiltration. He got the support of the two Associations in San Joaquin, the Independent Oil Producers Agency (representing producers at Kern River and Coalinga), and Fred Hillman, one of the men who spearheaded the development of the newly independent Standard of California. It appeared that a critical mass of medium and large sized operators had come to the opinion that state regulation was needed to control the many operators, some fly-by-night, who took shortcuts such as not sealing off abandoned wells.

In early 1915 a bill was passed to "protect oil sands menaced by water." This set up the Department of Petroleum and Gas within the State Mining Bureau, the forerunner of today's Division of Oil, Gas, and Geothermal Resources (DOGGR). Hamilton appointed McLaughlin as the first state oil and gas supervisor. One of the first things done was to require operators to keep records of their drilling operations. The Kern County and Coalinga Associations donated their records to start the collection. Companies that planned to drill a well had to file a "Notice of Intent to Drill," which had the well location, estimated depth of water shut-off, depth of oil or gas sands, and other data. They also had to keep log books detailing the work done by each twelve-hour tour of a drilling crew, including depth of the well, what formations had been encountered, whether oil or gas was present in the well, how much casing was put in or taken out, and more.

All of this record-keeping was meant to provide a "complete record or log of each well, giving in detail each and every step taken in its construction and repair, as well as the location and thickness of all strata penetrated so far as can be determined." Records were also kept of the production history of oil and water of each well and each field. All of this information is available today to the general public, and is invaluable for engineers and geologists who are exploring for new deposits of oil.

These well records played a central role in the development of petroleum exploration into an enterprise that uses the scientific method in its endeavors. Each medium and large operator began keeping copies of the public data, together with its private information, in a "vault" covering the regions in which it operated. Today this information is used to help determine where to drill in-field wells, design casing programs, well completions, well stimulation procedures and more. Geologists (company employees and independents) use the data to map geologic structures and estimate depths and locations of potential oil reservoirs adjacent to or near existing fields. Any clues, such as a change in drilling rate or density of drilling mud, or use of a new type of drilling bit, can be meaningful to an experienced geologist or engineer. Over the years oil seekers have supplemented DOGGR records in various ways, such as doing geophysical surveys or sending someone to observe oil field operations. Scouts have estimated how deep a competitor is drilling by using binoculars to count the number of stands of pipe when a driller is "tripping out." Any available clue is potentially useful, especially today in a state where most of the obvious pools of oil have already been found. DOGGR is currently in the process of uploading all of their old files online for public use through their webpage.

In addition to record keeping, the Department of Petroleum and Gas had enforcement power to require operators to follow procedures to shut off water, and to repair a well that was damaged. Such orders could be reviewed at the request of the operator by a board composed of fellow operators.

INDEPENDENTS: UNSEEN GIANTS OF CALIFORNIA'S OIL INDUSTRY

Large companies like Standard of California, Shell, Associated (affiliated with Southern Pacific Railroad) and Union are prominent in the annals of California oil, but small independent companies in their thousands have collectively made as big a contribution as any major company. This is especially true for many of the early, shallow-production fields such as Los Angeles City and Kern River. Independents participated in many of the innovations, including methods of controlling water intrusion in wells. We shall see in later chapters how independents have recently become the major players in California, getting the most out of mature fields with new technology.

Independents have been able to form associations to increase their influence on the industry. One such combination, the Independent Oil Producers Agency, was formed in 1904 at Kern River, largely as a result of a very low price of 11 2/3 cents per barrel offered by Pacific Coast Oil Company (then controlled by the Standard trust). Originally consisting of nineteen operators, the Independent Oil Producers Agency represented a significant portion of Kern River production. The agency leased the members' land and gave them licenses to operate their holdings. The Agency had the right to sell the oil and thus had considerably more negotiating power than any individual company. In 1907 the Coalinga Oil Producers Agency was formed by heavy oil producers in that field. In 1910 these two agencies merged.

Low oil prices plagued the independents in the early years of the century. The high production at Kern River and other fields was a major factor. Two major buyers were Standard of California and Associated. Many of the independents thought that these two colluded to keep prices low. By 1909 the two agencies accounted for more than 10 percent of California's production, according to contemporary estimates. Suddenly in May of that year the agencies made an agreement with Union to build a pipeline from San Joaquin Valley to the coast so that they would not have to use Standard's pipeline. Of this combination *California Oil World* said "For the first time in the history of oil in the West there is an absolutely dominant hand at the head of the business." Although this is an overstatement, it has a ring of truth. The combine was shortly to be joined by Associated. By 1919 the Independent Oil Producers Agency was responsible for 8.8 percent of California's production. Union and Associated each had almost the same production, for a total of 26.5 percent, surpassing that of Standard of California. Clearly, Standard of California had to rethink its policy of buying most of its crude from other producers. With a free hand since it had been released from its ties to the Standard Rockefeller trust in 1911, Standard developed its own exploration program and went on to become the giant we know today as Chevron.

The California independents went on to endure many ups and downs of the oil industry. Competition with majors continued, as it does today. Nearly all of the production at Kern River was eventually taken over by Chevron; somewhat ironically, the actions of those independents strengthened Chevron immeasurably. Most of the small, early companies are forgotten, but some still exist after a century or more. Younger independents have joined them to create the dynamic California industry of today, as we shall see in Chapter 5.

If necessary law courts could be called on to enforce orders. Companies were required to follow procedures when abandoning a well so that water could not travel through the well bore to a productive sand. The department initially had an annual budget of $45,000, out of which McLaughlin hired four assistants who were assigned to different geographic areas. They had to regulate an industry that had some 7,000 producing wells, almost 81,000 acres of proved oil land, 2,000 miles of pipelines, 30 refineries, and about 40 tanker ships. It was a Herculean task. Gradually a *modus operandi* developed in which operators cooperated to make the system work, because it was in the industry's interest. An important reason why this government regulation has worked is that it was devised over time with the participation of the industry being regulated, rather than being imposed "top down" without the expertise of those operating in the fields.

The Kern River Field went on to produce about 2 billion barrels as of 2007, third in California after Wilmington and Midway-Sunset, and fifth in the U.S. The field now has about 9,000 wells in an area of 17 square miles. Remaining reserves are probably around 500 million barrels and annual production was about 26 million in 2012. The early small companies were eventually consolidated so that only a few operators such as Tidewater, Getty, and Texaco were left. Finally these were for the most part bought out by Chevron, whose ancestor had built that original 8-inch pipeline. This made it possible to operate the field in a consistent, rational manner, using all of the log and other data to design ways to maximize production.

GEOLOGY COMES TO THE FOREFRONT

More discoveries were made in the first years of the new century, and Kern River in its turn was eclipsed. Drilling near seeps was still common. However, geological ideas about anticlines and faults increasingly became prime considerations in exploring for oil. At first this meant looking for outcrops and surface expressions such as lines of low hills. Companies varied in how much trust they placed in the new ideas. Geologists with their fancy degrees from Stanford or Berkeley often met resistance from hard-scrabble oil men who with their years of experience thought they could better manage a well without interference. Progress was slow but it was sure.

The Santa Maria Field near the present-day town of Orcutt in northern Santa Barbara County was opened by Western Union Oil Company, part of the Shell combine, in 1902.

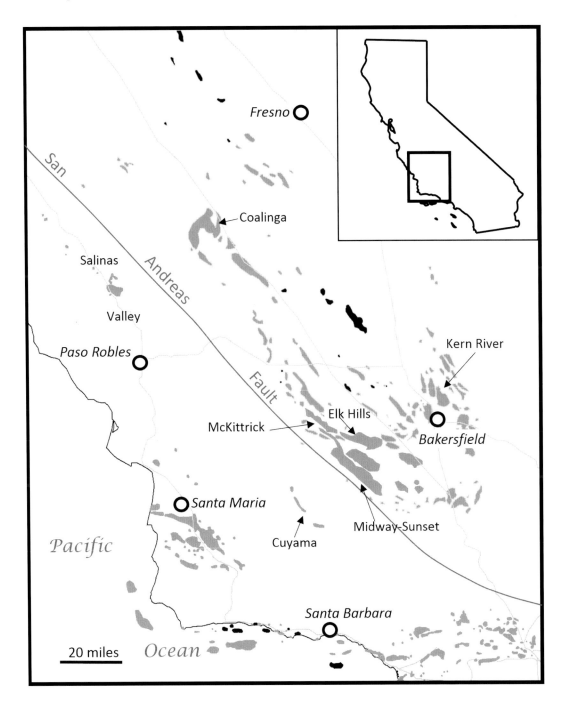

Map of present-day oil and gas fields. San Joaquin Valley is in the center-right portion of the map. Also shown are fields near Santa Maria (Orcutt Hill), Cuyama, Salinas Valley, and offshore Santa Maria and Santa Barbara, all of which are discussed in following pages. Oil fields are grey, gas fields are black. Present-day major highways are shown as grey lines.

Above: Members of the Hartnell family looking at the oil in the "Old Maud" sumps. Their faces suggest differing opinions about what they are seeing.

COURTESY OF DARWIN SAINZ FAMILY.

Right: "Old Maud" (Hartnell No. 1), 12,000 barrel a day gusher, 1904.

COURTESY OF DARWIN SAINZ FAMILY.

Below: Sumps, or lakes of oil from "Old Maud."

COURTESY OF DARWIN SAINZ FAMILY.

This drilling was based on outcrops of "asphaltum" on canyon walls. Other prominent operators in the area were Pinal and Dome Oil Companies, both owned by J. F. Goodwin and his partners. Union Oil, with new ideas and an aggressive leasing policy, would prove to be a major player. After sending its geologist, W. W. Orcutt, to survey the area, Lyman Stewart leased almost 70,000 acres. Striking oil only three days before the leases expired, Union went on to drill 38 wells. One of the most famous of all oil wells, "Old Maud," blew in at 12,000 barrels a day in 1904. It took three months to get the well

Union Oil's famous Gusher, Santa Maria, Cal.

under control. Water was a serious problem at Santa Maria, as elsewhere in California. Union's Drilling Superintendent Frank Hill came up with a bailer that had holes to allow cement to flow out of it and outside of the casing, sealing the well from higher water sands. This was one of the first down-hole cementing jobs attempted.

Above: Excavation of La Brea Tar Pits for fossils, by Orcutt and others, about 1900-1915. Oil derricks are visible just north of the tar pits. Bones up to 38,000 years old have been found. The modern day tar lake (left), complete with concrete models of trapped mammoths, was once an asphalt mine. Hancock Park is now surrounded by densely populated city, and no signs of the early oil field remain. The George C. Page Museum, at the park, has a million fossils from the pits, including mammoths, saber tooth cats, bisons, camels, and more.

EARLY PHOTO COURTESY OF STEVE MULQUEEN.
MODERN PHOTO BY AUTHOR.

Below: W. W. Orcutt, Union Oil's geologist.
COURTESY OF CALIFORNIA OIL MUSEUM,
SANTA PAULA, CALIFORNIA.

The town of Orcutt, named after the company geologist for the many oil discoveries credited to him, was a company town built to house the many workers needed for this successful oil field. Orcutt was considered the "dean" of California petroleum geologists. He established a petroleum geology department at Union and used surface geology to explore successfully for petroleum. He pioneered the use of aerial photography (from rickety early biplanes) to record surface geology. He made use of oil seeps, but he incorporated this information into his geologic mapping of potential oil-producing formations, rather than blindly drilling any location merely because it was near a seep

Top: Going up Hartnell grade, Santa Maria Basin, hauling equipment for a gas plant with seven horses, five mules and a steam traction engine.
COURTESY OF DARWIN SAINZ FAMILY.

Middle: Pinal No. 26, Orcutt Hill, Santa Maria Basin. Early internal combustion engine, with flywheels spinning, is driving the belt that goes into the shed at right.
COURTESY OF DARWIN SAINZ FAMILY.

Bottom: Working on tools, Pinal No. 26 well. A heavy cable tool is resting on the anvil. The tool dresser had to maintain and repair these tools using the forge in the shed next to a well.
COURTESY OF DARWIN SAINZ FAMILY.

Orcutt was a remote place in those days. As at Pico Canyon and many other oil fields, workers and their families lived close to the oil wells. Transportation was difficult, and the necessities of life had to be brought to the people who lived there. The hilly terrain made it difficult to bring in the heavy equipment needed. Draft animals and steam engines were sometimes combined to do the job. Electrical power was often lacking, and steam engines or early "hit-and-miss" internal combustion engines were used to generate electricity or directly supply motive power. Blacksmiths and other skilled artisans had to improvise solutions to problems, usually "lost tools," as they arose. In spite of remoteness the companies and their workers were able to build an infrastructure that included not only oil wells, but pipelines, processing facilities such as gas plants, tankage, and transport terminals.

Back in San Joaquin Valley, production at Oil City was in decline in 1900 after just a few years. Then in 1901 the Coalinga West pool was brought in. The Coalinga East pool followed about a year later. In 1909

the Silvertip gusher blew in, producing 102,000 barrels in a month. The Los Angeles Stock Exchange was closed for a day so that its traders could go to Coalinga and witness the gusher for themselves. A pipeline was built 110 miles west to Monterey on the coast, and a branch was built to the Standard Oil Kern River—Richmond pipeline.

Above: Brookshire Lease, Rice Ranch Oil Company. A truck from Orcutt Mercantile is delivering supplies.
COURTESY OF DARWIN SAINZ FAMILY.

Below: Life on Orcutt Hill, 1908.
COURTESY OF DARWIN SAINZ FAMILY.

Top: Pinal Gas plant, 1916.

Middle: Interior of Pinal Gas plant, 1916.

Bottom: Internal combustion engine in rear is driving a generator in the foreground, Pinal Gas plant, 1916.

The town of Coalinga boomed. Although the nearest railroad depot was twenty miles away, people got there somehow. The swelling town had a roughneck district, with "Whiskey Row" at the center. The streets were mud, and conveniences like baths were rare and expensive (for the second person using the bathwater it was cheaper). Gambling could be had anywhere. In 1913 much of the production of the area was purchased by Shell, giving that company a significant foothold in California. Shell built a company town called Oilfields near Coalinga. Almost a billion barrels have been produced at Coalinga, and remaining reserves are about 60 million. Coalinga, like several other fields along the western margin of San Joaquin Valley, is in an anticline that has surface expression as a low, elongated hill.

In 1889 preparations were made to drill near some tar seeps in the vicinity of present-day

Maricopa, southwest of Bakersfield. This would ultimately be recognized as another anticline, and the field would be named Midway-Sunset. Drilling activity continued at a slow pace through the 1890s and early 1900s. A small refinery was built at Maricopa. Daily production was less than a few thousand barrels until 1908. Oil field workers and their families lived among the derricks in towns like Taft and Sunset. Higher oil prices stimulated drilling, and daily production rose to almost 150,000 barrels in 1914. Eventually, 22 individual oil reservoirs were found in six formations in the 30 square mile area of the field. Discoveries were still being made in the 1980s. By 2008 the field had more than

Above: Auto race through the streets of Oilfield on Washington's Birthday, 1912.
PHOTO FROM BILL RINTOUL COURTESY OF GENERAL PRODUCTION SERVICES.

Left: "Derrick Blvd." at Coalinga, an example of evenly spaced wells, a more efficient way of developing a field than the chaotic town lot drilling of several of the early fields.
CALIFORNIA STATE LIBRARY.

Below: Monarch Maricopa Refinery, Midway-Sunset Oil Field, 1907.
The Temblor Range is in the background.
PHOTO FROM BILL RINTOUL COURTESY OF GENERAL PRODUCTION SERVICES.

INCORPORATED UNDER THE LAWS OF THE STATE OF CALIFORNIA
DECEMBER 9, 1908.

NUMBER
A236

SHARES
42,250

LAKE VIEW OIL COMPANY

CAPITAL STOCK $2,500,000.
2,500,000 SHARES OF $1.00 EACH.

THIS CERTIFIES THAT **GRACE M. OFF** * * * * * * * * * * * * * IS THE OWNER OF

* * * FORTY TWO THOUSAND, TWO HUNDRED FIFTY * * * * SHARES OF THE CAPITAL STOCK

OF THE LAKE VIEW OIL COMPANY

TRANSFERABLE ONLY ON THE BOOKS OF THE COMPANY BY THE HOLDER HEREOF IN PERSON, OR
BY ATTORNEY ON SURRENDER OF THIS CERTIFICATE PROPERLY ENDORSED.

IN WITNESS WHEREOF, THE SAID COMPANY HAS CAUSED THIS
CERTIFICATE TO BE SIGNED BY ITS DULY AUTHORIZED OFFICERS, AND TO
BE SEALED WITH THE SEAL OF THE COMPANY THIS ___13TH___ DAY
OF ___DECEMBER___ 1916.

John McGrath
SECRETARY

W. W. Orcutt
PRESIDENT

Off and Fried made a deal with a much larger company that controlled a neighboring piece of land. This was Union Oil, which seems to have a piece of many of the big stories of California oil. Union wanted Lake View's land for a tank farm, not to drill for oil. The deal gave Union 51 percent of Lake View's stock in exchange for drilling the well when their crews were not busy drilling their own wells. Union put one of its drillers, Charles Woods, whose unfortunate nickname was "Dry Hole Charlie," on the job. Charlie had more than a dozen dry holes for Union, but when he hit one, he really hit it. The well gushed 125,000 barrels the first day. This was more than the whole rest of the field

Top: In the early days gushers were signs of success (not mistakes) and were great draws for spectators. Here, the San Francisco Stock Exchange is visiting the Lake View gusher in its dying days. They brought their flag and carefully posed for the occasion. Hopefully they did not get those white shirts and coats dirty.

LIBRARY OF CONGRESS.

Above: Stock certificate for Grace M. Off, dated in 1916. This was six years after the Lake View gusher, when the company was controlled by Union Oil. Union's geologist, W. W. Orcutt signed the certificate as president of the company.

COURTESY OF THE OFF FAMILY/OJAI OIL COMPANY.

Right: Crater left by the Lake View gusher. A small amount of oil is still coming out. Men working in the crater are soaked with oil.

PHOTO FROM BILL RINTOUL COURTESY OF
GENERAL PRODUCTION SERVICES.

produced. A month later it was still making 90,000 a day. They had to build earthen levees to contain the flow. As it turned out less than half of the oil was saved. As for Dry Hole, he went on to drill another dozen dusters.

Although these great gushers and the oil fields they birthed were often discovered as the result of seeps, or perhaps ideas about red grass, they became the proving grounds for solid geological principles that would make for still greater discoveries. Data from well logs and records housed at the Department of Petroleum and Gas made it possible to map the anticlines in west San Joaquin Valley. The anticlines are more or less parallel to the San Andreas Fault, which runs just to the west, close to the boundary between the Valley and the Coast Ranges. The association of anticlines with a fault is something that has been repeated time and again in California.

That the movement of a fault can cause folding of the crust into an anticlinal oil trap is a scientific explanation that can lead to very sophisticated ways of looking for oil. For example, knowledge of when in geologic time a fault moved can imply when an oil trap formed, and whether oil was available at that time to migrate into the trap.

Los Angeles Basin was to become the stage where the new ideas succeeded spectacularly. Two of the largest pre-1900 discoveries in the basin, the Brea-Olinda and Whittier Fields, are in a range of prominent hills. Although these fields had been discovered by drilling near seeps, development of the fields revealed that they were contained in an anticline that ran along the range of hills. Another range with smaller hills, including Coyote Hills, lay to the southwest of and parallel to the larger range. Several more fields were discovered

Below: Oil fields in Los Angeles Basin discovered between 1900 and 1919 (labeled). Most of these fields are in a range of low hills parallel to and about five miles southwest of the range where the earlier discoveries were made.

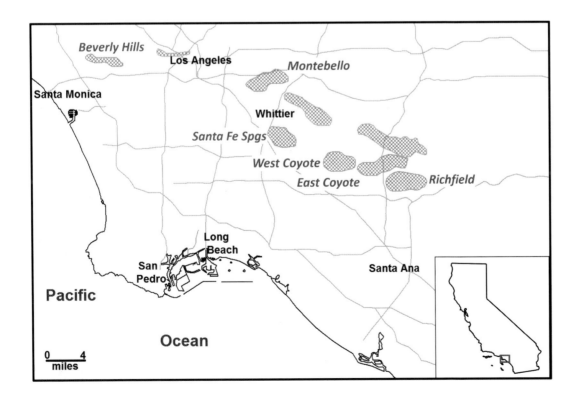

Right: Santa Fe Springs Field in 1929.

Below: Beverly Hills Field, discovered by W. W. Orcutt in 1900 by drilling on a slight ridge or spur extending from the Santa Monica Mountains to the north. The La Brea Tar Pits are located about a mile to the west. This view is looking west from Highland Avenue just south of Sixth Street. Unlike the Los Angeles City Field, this area was mostly rural at the time. Wells continued to be drilled in this field, right up to the present, as the city grew up over it.

there between 1900 and 1919 using geological ideas about anticlines rather than simply drilling near seeps. The West Coyote Field was discovered in 1909 by drilling on a broad, 300-foot high hill. Geologist William Plotts of Murphy Oil predicted oil would be there, perhaps because a local water well had a show of oil, but the anticline idea inspired drilling on the hill.

Early development at West Coyote revealed the outline of the anticline, which led to the discovery of the other fields in the trend. First came East Coyote in 1909.

Next were Richfield and the giant Santa Fe Springs Field, the latter six miles to the northwest of West Coyote. Santa Fe Springs was in a seemingly flat area. Nevertheless it is said that J. Paul Getty's father George F. noticed a freight train straining as it traveled the apparently flat land, then suddenly begin to speed up even as it ceased to labor, indicating it had reached the top of an almost imperceptible hill. The Gettys leased four small lots and proceeded to drill the Nordstrom No. 1, which was a 2,300 barrel a day producer. Although this was not the

field's discovery well it highlighted the importance of geologic structures that sometimes show the minutest possible surface manifestations. Another discovery was the Montebello Field in 1917, which was on the Whittier-Brea-Olinda trend. These fields, with their anticlines and surface expressions, spurred wildcatters to look for similar surface structural expressions without reference to seeps, leading in the early 1920s to a quick succession of discoveries on yet another trend that added more than 3 billion barrels of oil to California's reserves.

FROM CABLE TO ROTARY

Until after 1920 virtually all wells in California were drilled using cable tools. In this method a heavy bit was suspended on a rope (usually hemp in the early, shallow wells and wire rope for deeper wells) that was raised and lowered, causing the bit to repeatedly pound at the bottom of the hole. Periodically the bit was removed and a bailer, a long, heavy pipe, was sent to the bottom to remove the cuttings and other debris from the hole. The bailer had a door on the lower end

Below: Cable tool drilling rig.
A steam engine provided power to the band wheel via a belt. The band wheel caused the walking beam to rock up and down on the appropriately named Samson post. This imparted the up and down strokes to the drill string.
COURTESY OF STEVE MULQUEEN.

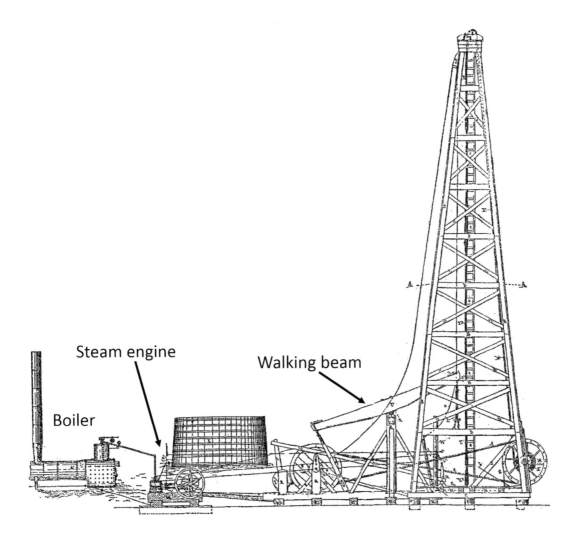

Steam engine

Walking beam

Boiler

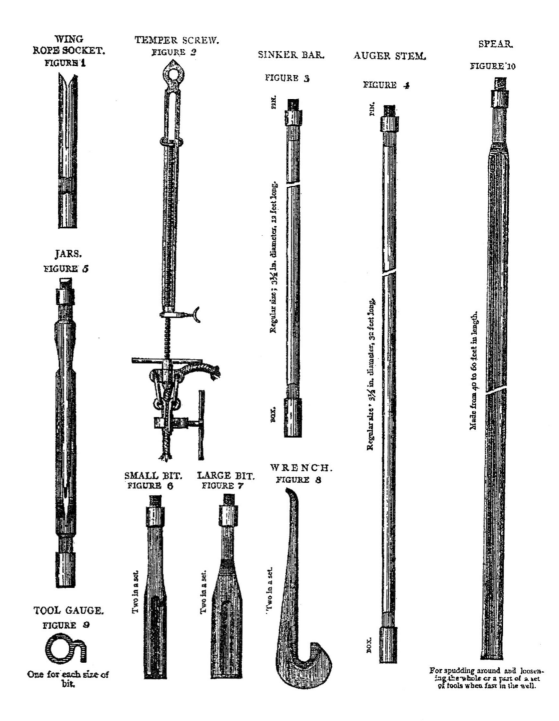

WING
ROPE SOCKET.
FIGURE 1

TEMPER SCREW.
FIGURE 2

SINKER BAR.
FIGURE 3

AUGER STEM.
FIGURE 4

SPEAR.
FIGURE 10

JARS.
FIGURE 5

SMALL BIT.
FIGURE 6

LARGE BIT.
FIGURE 7

WRENCH.
FIGURE 8

TOOL GAUGE.
FIGURE 9

One for each size of bit.

For spudding around and loosening the whole or a part of a set of tools when fast in the well.

Right: Cable tools. Bits were shaped like a chisel or the end of a fish's tail. Jars were joints in which the upper part could be lifted up to thirteen inches without lifting the lower part; this was used to cause a sudden, sharp blow on the upstroke to dislodge a stuck set of tools. The auger stem was a very heavy bar, thirty feet long and weighing half a ton. This weight provided the inertia for the downstroke of the bit. A spear was just one of hundreds of tools for "fishing," the process of recovering tools, casing, or other objects that dropped or were stuck in the hole. Fishing tools were often improvised by the crew to solve a particular problem. Successful fishing tools, with names like alligator grab, collar grab and rope knife, made it into supply catalogues.
COURTESY OF STEVE MULQUEEN.

Opposite, top: Rotary rig. The drive chain, gear set and rotary table can be seen. This rig seems to be coated with mud.
COURTESY OF CALIFORNIA OIL MUSEUM, SANTA PAULA, CALIFORNIA.

Opposite, bottom: Base of a wooden derrick with its crew. The massive block is visible. Other derricks, two behind and one to the left, are quite nearby, indicating that this is probably town lot drilling. The men are pausing from their hard labor, with gloves in hand and tools close by, to pose for this picture. Presumably the slighter-built man at left in cleaner clothing is the boss. Shifts, or "tours" in those days lasted for twelve hours of relentless, dangerous work. No hard hats back then. These nine men represent the backbone of the petroleum industry. Without them it would be nothing.
COURTESY OF HISTORICAL SOCIETY OF LONG BEACH.

that would open to accept the cuttings, and then close as the bailer was pulled back up. The wooden derricks were about 50 feet high, and power was usually provided by a steam engine. A small crew of only two was needed, the driller and a tool dresser. The tool dresser kept the bits and other equipment in repair using a forge next to the derrick. Tool dressers were expert blacksmiths, and were often ingenious inventors of whatever tools were needed in any situation, such as "fishing" out a stuck drill string from the hole.

Rotary drilling is a much more complex operation than cable tool drilling, although in

principle it is no more complicated than a household electric drill. The cable is replaced by a string of pipe with the bit at the end. Mud circulating down the pipe and up via the annulus carries the cuttings out of the hole. A mud cake against the wall of the hole keeps the mud from disappearing into the rock and helps to prevent the hole from caving in. "Tripping out" is a time-consuming operation in which the pipe is stacked up in the derrick, often three joints at a time, meaning that the derrick has to be about 100 feet high. Crews consist of five men, the driller, three roughnecks on the derrick floor, and the derrick

man who goes up to the top of the derrick to handle the stands of pipe.

Rotary drilling became popular along the Gulf Coast of Texas after 1895. By the Spindletop discovery in 1901, more than 100 wells had been drilled with rotary tools there. Although rotary drilling was tried in California in 1889 at Pico Canyon, as we have seen, the experiment was less than successful because the rotating bit could not grind through hard cobbles. In 1902 M. K. Oil Company drilled a 2,400 foot hole in Coalinga with rotary tools, but the hole was so crooked that they could not run casing. Amalgamated Oil rotary drilled a

2,357 dry hole near the Salt Lake Field in Los Angeles at about the same time. In 1908 Standard Oil of California decided to import some rotary rigs and crew from Louisiana to drill wells in the San Joaquin Valley. The same problem with cobbles kept coming up. While a cable tool could pound its way through a cobble, a rotary bit would turn on top of the cobble while fine material was scoured out by the circulating mud. The cavity so formed would then fill with other cobbles, creating a barrier that the bit could not penetrate. This cobble problem was quite severe in many California oil fields, while on the Gulf Coast sands and shales could easily be drilled with rotary tools.

Other problems were less real and may have existed mainly in the minds of those who resisted change. One of these was the idea that the mud cake would seal off an oil sand, preventing oil from entering the bore hole. Because of this some operators started a hole with rotary tools and switched to cable just before reaching the projected oil sand. There was also prejudice and rivalry in cable tool crews, who called the rotary crews "swivel necks."

In 1910 the first gusher drilled with a rotary rig in California came in at 1,500 barrels a day at 2,432 feet (Standard Oil). After that the technology slowly improved and adapted to California conditions. By 1911 Union Tool Company (one of Lyman Stewart's brain children) was manufacturing rotary rigs so they no longer needed to be brought in from out of state. Union Tool also came up with a twin steam engine in which the two pistons were attached to the crankshaft a quarter turn out of phase so that the engine could be restarted easily from any stopped position. Guards were put over the rotary tables' power chains, which were extremely dangerous if they broke. In the

1920s a driller in Louisiana got the idea of mixing dense minerals with the mud to make it heavier in order to avoid blowouts. Soon barite was being used everywhere, including in California. The wooden derricks were gradually replaced with steel ones in the 1930s. Also in the 1920s the grip ring assembly, by which the pipe was gripped by the rotary table, was replaced by the kelly, a square cross-section length of pipe that fit into the kelly bushing set in the table. The fish-tail drill bit, a leftover from cable days, was finally replaced by the new rolling-cutter bits.

Rotary drilling goes a long way toward solving a major problem of cable tool drilling: gushers. Although gushers were heralded as signs of success in the early days, they resulted in wastage of much of the produced oil and in a loss of formation pressure, so that most of the oil in the reservoir could never be produced. When a cable tool penetrates a pressured formation there is nothing to prevent the fluids from "blowing out." On the other hand, the drilling mud used with rotary drilling can be made heavy enough to prevent this, if the overpressured formation is anticipated.

Opposite page, rotary rig floors and the crews who man them:

Top, left: The man who is leaning with his hand just below the rotary swivel looks like he is in charge; at least that is what he wants you to think. Below the swivel is the kelly, which is square in cross section and fits into the kelly bushing in the round table below.

Top, right: This picture clearly shows the square kelly. This same picture shows the chain that drives the rotary table, with a safety cover. The roughneck leaning over the safety cover seems to be saying "It can't get me now."

Bottom, left: A chain in another probably earlier picture is exposed. Most of the five men are smiling but the younger man closest to the chain looks a little dubious. These chains could easily take an arm off, and the safety cover was added for good reason.

Bottom, right: Six other men are gathered around two drilling bits, enjoying their cigars. The cross-shaped bit could have been used in a cable tool rig. The other bit is a roller type designed for rotary drilling.

THE FOUR PHOTOGRAPHS ARE COURTESY OF THE HISTORICAL SOCIETY OF LONG BEACH.

Left: Jack line cables running from offset wheel. As this wheel rotates the cables are pulled back and forth, each running the pumpjack on a well. Thus several wells could be pumped using a single steam or hit-and-miss engine.

COURTESY OF CALIFORNIA OIL MUSEUM, SANTA PAULA, CALIFORNIA.

BUSY CABLES & BLACK GOLD

Opposite: Transition. A steel and a wooden derrick side by side. Taken from Signal Hill. The sun is setting over Palos Verdes Hills.

Left: Steel derrick in the Brea-Olinda Field. These replaced the old wooden derricks, but they in turn have been replaced by portable rigs. Nearly all of the thousands that once dotted the landscape of southern California have been removed. This is one of a mere handful left standing in 2015. At least one other has been repurposed as a cell phone tower.

Above: Newspaper advertisement for Red Crown Gasoline, about 1917. Purity of gasoline was an important advertising point. Standard claimed that their "straight distilled" gasoline, not a mixture, gives easy starting, quick acceleration, and good mileage. The ad uses a teakettle as an analogy for the "continuous range of boiling points" of their fuel.

Below: Delivery truck in Oakland.
COURTESY OF CHEVRON USA.

THE AUTOMOTIVE AGE ARRIVES

In the first two decades of the twentieth century the uses of petroleum changed dramatically. Previously, kerosene for illumination, fuel oil, and lubricants were the mainstays. However, the new era of transportation brought gasoline to the forefront. By 1910 the amount of gasoline manufactured exceeded that of kerosene. By 1920 the little town of Los Angeles had become a metropolis of 577,000, and many other cities dotted the landscape of southern California, linked by the rails of the Pacific Electric and by a continually improving network of roads. Unlike cities back east, these towns grew during the dawning age of the automobile, and they spread out. From 1910 to 1920 automobile ownership went from 1 for every 75 people to 1 for every 7. The southern California car craze was on, and it needed one thing more than anything else—gasoline.

Marketing operations of most major oil companies before 1910 were mostly wholesale. For example, Standard of California, which dominated the market for all products except fuel oil, had "stations" (really district warehouses) that used horse-drawn wagons and later trucks to deliver products in tanks and barrels to retailers. Distillate, a fuel heavier than gasoline, was delivered to factories for use in stationary engines. Kerosene, and later gasoline, were usually sold to consumers in tins. Before about 1913 a motorist would go to a hardware store or even a grocery store and purchase gasoline in five-gallon tins to store at home for later use. At Union Oil's yard at Sixth and Santa Fe in Los Angeles, a row of fifty-gallon tanks was set up, each with the name of an individual motorist. When the car owner wanted some gas he simply drove to the yard and filled up himself. On a trip one would seek out a garage, a car dealership, or even a bicycle shop that might be selling gasoline as a side business. The gas would be poured from a barrel into a measuring container and then into the car's tank. Sometimes a small buggy with a tank and a hose was used to gas up a car pulled to the curb of a street, but this caused traffic problems. Quality varied, and merchants sometimes adulterated the gas with the cheaper and less desirable distillate. As the Model Ts and Chevies proliferated, this had to change.

In 1907 a Standard Oil warehouseman in Seattle repurposed a water tank, fitting it with a hose and a valve, and sold gasoline to customers who would come to the main plant. Thus Standard lays claim to having the first "service station" anywhere. The next move was to have the service station by itself, away from the warehouse and in every neighborhood. Such stations began to appear in Seattle and elsewhere. In 1912, Standard tried to build four neighborhood stations in Los Angeles. Citizens, perhaps encouraged by garage owners, objected on grounds of fire hazard and possible damage to their property values, and got the city council to deny the permits. In the following year several auto dealers including Earl C. Anthony, later owner of the famous KFI radio station, formed National Supply Stations, Inc. They built a little one-pump gas station with a 12 by 15 foot wooden building whose roof extended over the pump. They sold Standard's Red Crown gasoline, and had a sign with the red, white, and blue colors still used by Chevron today. In a few months they had nine stations in Los Angeles and Pasadena, and in January

1914 accounted for 22 percent of Los Angeles gasoline sales.

Not to be outdone by Standard, other companies were getting into the act. Union Oil opened its first station in downtown Los Angeles in 1913. By 1914 stations were beginning to appear in places like San Francisco, San Diego, Berkeley, Santa Ana, Oakland, and Anaheim. By the end of 1914 there were fifty stations in Los Angeles. Stations appeared in the Sacramento and San Joaquin Valleys, and California oil companies supplied stations in Oregon and Washington. The "gas pumps," with glass bowls at the top graduated in gallons to accurately and conveniently dispense the fuel, were a big advance.

Companies touted their brands, such as Red Crown, Pinal Dome's Pennant brand, Motor Maid, Owl, and others. Purity and consistency were main selling points. Garage men and others competed desperately, but the convenience of the gas stations and the efficiency of their operation meant that gas stations would take over the market. Eventually service stations offered lubricants and conveniences like compressed air for tires, and even promotional items like road maps.

Another well-known feature of service station retailing emerged in 1914: the gas war. An oversupply of crude oil and refined stocks led to prices being cut from about 19 cents a gallon in 1912 to as low as 10 cents in 1915.

Above: Garage selling gasoline. The brand is "Aeroplane." The garage also repairs batteries, sells tires, has a machine shop, services both cars and tractors, and advertises a ladies' rest room. Two doors down is a horse shoeing establishment.
COURTESY HISTORICAL SOCIETY OF LONG BEACH.

Below: Hardware store selling gasoline.
COURTESY OF CHEVRON USA.

Early gas station operators (pages 58-61) experimented with different ways to market their products. Some sold only one brand of gasoline, others sold as many as four or five. One photo shows a station with five pumps lined up, each painted differently to represent a brand. Stations offered restrooms, tires, oil, and S&H Green Stamps (opposite, bottom). Uniformed attendants wearing bow ties (opposite, top) would stand by ready to serve. Some stations had six or more islands. By the '40s the glass bowl pumps(opposite, bottom) were mostly gone. A Richfield gas station in Long Beach (below) was also an Oldsmobile Dealer. The pumps, where an attendant is shown airing up a customer's tire, are just outside the showroom. While getting gas for his beat-up old car the customer could look longingly through the window ten feet away at the spiffy new Oldsmobile of his dreams.

ALL PHOTOGRAPHS COURTESY OF HISTORICAL SOCIETY OF LONG BEACH.

Finally prices began to stabilize in late 1915, ending the first of many gas wars to come. Suppliers often gave 2 percent discounts to gas stations that paid in cash, and also lent equipment like tanks. Rather than discounting, Standard instead chose to market its Red Crown as being superior to other gasolines in starting, acceleration, and fuel economy.

Late in 1914, Standard purchased National Supply and its 31 stations. New stations were built of steel and were landscaped with flower beds. Attendants wore white uniforms and were expected to take care of the shrubbery when they were not busy with customers. By 1919 Standard led the field by far with 218 stations. Associated had 85, Shell, a new

entrant in the California petroleum industry, had 77, and Union had 32.

Union Oil did not initially get into service stations in a big way like Standard, but they soon made up for lost time. Union had a much larger market share of fuel oil than Standard and lagged in refined products. Their first gas station at Sixth and Mateo in Los Angeles was set up only because motorists lined up around the block at their yard. Union's entry into the retail market was still quite meager until it acquired the Pinal-Dome Oil Company. Pinal-Dome was a small concern that produced mainly in Santa Maria Valley and had 20 service stations mostly in the Los Angeles area. Pinal-Dome had been

forced into the retail market earlier as a means of dispensing their production. With a less aggressive exploration program than Union's, Pinal-Dome saw its production fall to the point that it could no longer fully supply its retail outlets. Union needed more outlets for its production and the merger was thus a perfect fit.

Almost at the same time that Union got Pinal-Dome's 20 gas stations, its newly built Wilmington Refinery began churning out gasoline. Union quickly built more gas stations, putting on a contest among architects to come up with a functional, attractive design. Business skyrocketed as the auto boom really got under way. By the early 1920s Union was issuing credit cards to its customers. Union eventually developed a system of leasing the stations out to independent small business people, keeping only a handful in company ownership for training and to try out new ideas. The independent operators and their employees were trained to provide a consistent type of service and to maintain the stations in a clean, attractive condition. Union needed a name to market its high-octane anti-knock gasoline as a counter to the gas wars, which became potentially more ruinous in the depression years of the 1930s. Robert D. Matthews, a vice president specializing in accounting and finance, was a Welsh immigrant who was studying for his citizenship exam. He proposed the patriotic symbol "76," which fortuitously was also the octane value of the anti-knock gasoline coming out of the Union refineries. The name stuck, and it is still here today even though Union Oil no longer exists and the gasoline is now being sold by Tosco (Phillips Petroleum).

Gas stations grew in size and complexity. Most of them added service bays where major repair work could be done, thus ironically recalling the early days of garages which had lost their gasoline business to the stations. Multiple islands served many cars at once. Some of us can recall when we could drive into a gas station and just sit there while the white-attired attendant would check our oil and wash our windows with those blue towels while we listened to the "ding-ding" of our tank being effortlessly filled. We could get enough complimentary cups and plates to fill a whole table if we kept coming back. For a while there were "full-serve islands" and "self-serve islands," or just "full" and "self." Not any more, except in a few places like Oregon. Now the service bays are mostly gone. Today we can get anything we want for our stomachs and our cupholders. One thing is still the same: we have a delivery system for fuel for our vehicles that is so efficient and effortless that we completely

take it for granted. Of course, once in a while that is not the case, as in the early 1970s when there was gas rationing in response to the OPEC oil embargo, or the now forgotten episode in 1920 when an acute shortage forced Union and other companies to order special trains of gasoline-filled tank cars to come from Texas. Such events, however, are the exception and not the rule.

Above: Wells along the beach, Huntington Beach.
COURTESY OF CHEVRON USA.

Right: Inland part of the Huntington Beach Field. Downtown Huntington Beach and the pier are in the foreground.
LOS ANGELES PUBLIC LIBRARY.

Below: Signal Hill, part of the Long Beach Field.
COURTESY OF CALIFORNIA OIL MUSEUM, SANTA PAULA, CALIFORNIA.

Opposite, top: Long Beach Field skyline showing a seemingly impenetrable curtain of derricks. Signal Hill is on the left. Photograph probably taken in the 1920s or 1930s.
COURTESY OF HISTORICAL SOCIETY OF LONG BEACH.

Opposite bottom, left: Union L. B. C. No. 11 gusher. The spectacle has drawn a crowd.
COURTESY OF HISTORICAL SOCIETY OF LONG BEACH.

Opposite bottom, right: Shell-Martin No. 1 gasser, November 17, 1921. Men in the far right foreground are preparing their equipment in an effort to stop the flow.
COURTESY OF HISTORICAL SOCIETY OF LONG BEACH.

lots so that they could collect royalties on their new-found oil. There was also money for lawyers who fought over who owned the land where the phantom streets were supposed to be. Today the lots are largely undeveloped scrubland used for horseback riding. Some of the wells are still producing.

The next giant field along the trend was the Long Beach Field, which includes Signal Hill. Although exploration there started as early as 1916, oil was not discovered until 1921, slightly after the Huntington Beach initial find. Before the U.S. gained ownership of California, Signal Hill, north of what is now Long Beach, was known as *El Cerrito*, or "Little Hill." Used as a lookout and signal post probably for centuries, it acquired the moniker "Signal Hill" after a triangulation marker was placed on the summit in 1889. The original ranchos on Signal Hill, Los Alamitos and Los Cerritos, were subdivided into town lots before oil was found.

In 1916 Union Oil drilled a well at Wardlow Road and Long Beach Boulevard, just northwest of Signal Hill. They abandoned the hole at 3,449 feet, not knowing that the rich oil-bearing sands of the Signal Hill reservoir lie slightly deeper. Had they continued drilling the story of Signal Hill might have been very different. In 1919 geologists with Standard of California reported Signal Hill as "reflecting an anticlinal folding or dome

knocking, offering small amounts of money to buy their lots. While most sold, one or two paid their back taxes and reclaimed their

Long Beach Field in the 1920s. Sign in front of the tent says "Parkford's Signal Hill Oil Syndicate No.1." Benches are piled up outside the tent. A group of well-dressed men and women is standing at the entrance to the tent. Several cars are parked nearby. It appears that some kind of meeting for potential investors has just concluded. The field is crowded with wooden derricks, tanks, steam boilers, stockpiled supplies, a welder's shop and other oil drilling paraphernalia. Note the stand of pipe in the derrick behind the tent.

which gives favorable indications of oil production." This statement shows the new thinking that placed geological structures over seeps as indicators of oil. Although Standard, now independent of the old Rockefeller combine, was one of the more forward-thinking companies, it did nothing with the recommendation because it had a policy of avoiding town lot plays.

In 1918, Frank Hayes, a geologist with Royal Dutch Shell, recommended that the company explore in Signal Hill. Company management, saying that several dry holes had been drilled on the similar Dominguez structure to the northwest, rejected the idea. Changes in management led to this decision being reversed two years later, and Shell leased 240 acres on the east slope of the hill

WINSTEAD PHOTO - L.B.

that were deemed unsuitable for homes. Shell did not obtain all of the available leases on the top of the hill, and other companies passed on the opportunity. An executive with Union Oil, reflecting his company's earlier failure, said he would "drink every drop of oil" that Shell would find at Signal Hill. As it turned out it was a good thing he did not shake hands on this deal.

The Alamitos No. 1, at the intersection of Hill and Temple Streets, was spudded by Shell on March 23, 1921. Drilling with a rotary rig to 2,765 feet, they cored an oil sand. Shell leased more land. The word was getting out that something was up. Setting casing, they switched to a cable rig because of the then common fear that the rotary mud cake would seal off the producing formation. At 3,114 feet

Top: Long Beach Field, with a wooden derrick under construction. A group of men are conversing in front of the derrick. The lot next to the derrick is for sale.

What appears to be a gas plant for extracting casinghead gasoline is in the left background. All of the derricks in this scene are wooden, probably placing the picture in the 1920s.

COURTESY OF HISTORICAL SOCIETY OF LONG BEACH.

Middle: Long Beach Field, flat area north of Signal Hill. A few steel derricks appear in the forest of wood. This picture was taken a bit later.

COURTESY OF HERLEY FAMILY.

Bottom: Field office of Shell, which drilled the discovery well at Signal Hill. This photograph was taken in 1925, four years after the discovery.

COURTESY OF CALIFORNIA OIL MUSEUM, SANTA PAULA, CALIFORNIA.

BLACK GOLD IN CALIFORNIA: *The Story of the California Petroleum Industry*

on June 23, a gusher of oil blew over the crown block and up to 114 feet in the air. The well quickly sanded up, and after cleaning up, oil started flowing to tanks. By now there were 500 spectators on hand to witness the sight. Shell had to erect a barricade around the well, which soon was producing 1,200 barrels a day.

The great Signal Hill boom was on. No single company, including Shell, possessed more than a small fraction of the land, which was mostly in town lots. Derricks were sprouting like weeds, sometimes so close together that their legs supposedly interlaced. It was much like the Los Angeles City Field of thirty years before. It is said that one drilling crew fishing for a lost tool string snagged that of a neighbor whose hole had wandered over the property line. Automobiles bearing curious onlookers jammed the roads. Tents were set up where gullible investors were pitched while they ate chicken drumsticks. Some people's interesting concepts of arithmetic got in the way. One lot owner, offered a 1/10 royalty, refused thinking he could hold out for 1/20. Some hucksters erected fake derricks to make people think they were actually producing oil.

While Signal Hill was booming a lemon and avocado grower named Samuel B. Mosher was quietly running his farm in Pico Rivera, about ten miles north of Signal Hill. The Alamitos No.1 gusher would have been plainly visible from the windows of his farm house, as would the apocalyptic vision of dozens of pipes flaring gas every night, their glow perhaps giving Signal Hill the appearance of a man-made volcano. But Mosher had no time to play the tourist. The twenty-eight year-old was busy trying to feed his young wife and baby, often putting in twelve hour days on his tractor. Mosher's dream was to be a farmer. After graduating from UC Berkeley with a degree in agriculture in 1916, he spent five years developing his seventeen acre lemon farm, never earning more than $50 a month in his first two years. Finally, in the winter of 1922 he had his first big crop of lemons ready to harvest, bringing him hope of paying off part of his heavy mortgage.

Mosher had a fraternity brother named Robert Bering, who had gone on to become a geologist's assistant with an oil company. Bering told his friend what was going on at Signal Hill, arousing the curiosity of the young farmer, and perhaps infecting him with that "oil bug." Finally, on his twenty-ninth birthday, Mosher decided to take a day off and go find out what the commotion was all about. He drove his battered old Buick up to the lower end of Hill Street and walked up to the Alamitos No. 1 Well. Coming from the quiet of Pico Rivera, the noise of the oil field must have been deafening, especially with the roar of casinghead gas escaping into the atmosphere at high pressure from every well. This gas was regarded as a waste product at Signal Hill, not worth bothering with when the most important objective was to produce oil as quickly as possible.

The wet gas at Signal Hill contained a fraction that if separated from the rest of the gas would be a liquid, gasoline. Called natural, or casinghead gasoline to differentiate it from gasoline produced from oil at a refinery, this liquid could be blended with the refinery gasoline to create a high-test, or "ethyl" fuel suitable for high-performance engines. Mosher and Bering thought that if they could build a small plant to extract the natural gasoline, they could create a marketable product from the Signal Hill producers' waste gas. They made an informal agreement that Bering would set about securing leases for gas at Signal Hill while Mosher came up with the funds to build a plant. Mosher thought that his bumper crop of lemons, the first in his five years on the farm, would provide the capital to move the project forward.

Then disaster struck, the kind of disaster that sets in motion a series of events leading to something far bigger and better than what was originally hoped for. The main enemy of citrus growers in southern California is freezing weather. On January 19, 1922, a rare freeze brought the temperature at Mosher's farm down to nineteen degrees. Working all night setting out smudge pots, he saved his trees. However the entire crop of lemons and avocados was lost. Suddenly Mosher had no money for a natural gasoline plant, and worse, he had no money to pay his mortgage or buy groceries. After such a calamity many people would give up, but not Sam Mosher. He decided to forge ahead with the natural gasoline project, come what may. He sent off a penny postcard to request a free publication of the Department of the Interior, Petroleum Technology Pamphlet No. 176. This pamphlet showed, with diagrams, how to build for $4,000 a small plant using the absorption process to remove gasoline from natural gas.

Mosher in his optimism thought that all he needed to go into the gasoline business was $4,000. How was he going to get it? He decided to do the only thing he could think of, swallow his pride and ask his father for a loan. Henry M. Mosher was not predisposed to loan money for an oil venture, even to his son. Some twenty years earlier, he and other investors formed a company that drilled several producers in the Santa Maria Basin. Unfortunately the price of oil dropped, and they shut in their wells. When the price increased they opened their wells again, only to find they had gone to water. A neighboring, much larger company with tankage to store oil had in the meantime used offset wells to drain the oil from the reservoir. Sam explained in detail what he and Bering were planning to do and why it would be successful. However, Dad was not believing. Although Mrs. Mosher was sitting nearby apparently engrossed in a book, she was actually listening to every word passing between father and son. Finally, when H. M. flatly refused to loan the money, she said, "Dad, Sam will be getting the money after we're gone anyway. But he needs it now. If you don't lend him that $4,000, I will." Dad knew when he was beaten, and reached for his checkbook.

It turned out to be more difficult than they thought. For one thing, it was almost impossible to get producers to sign gas leases. Even though they would be turning over a waste product and would receive in return the residue gas for free for their boilers, and a 1/3 royalty on the gasoline, they did not want to sign life-of-the-field leases. After all, Mosher and Bering were unknown farm boys who as yet had no plant. Maybe the refiners would not buy the natural gasoline. Finally Bering said he had a contract with San Martinez Oil Company, his former employer. Based on this they got a surface lease on two acres for their plant. Mosher started looking into components that he would need to build the plant. These were in short supply, and much more expensive than originally thought. The government's three-year-old estimate of $4,000 was barely one tenth of what was actually needed.

Fortunately for Mosher, the giddy success of Signal Hill and the onset of the automobile age resulted in a climate of almost unlimited credit for oil ventures. He purchased equipment and supplies using miniscule down payments and easy terms such as no payments for ninety days, by which time Mosher and Bering

were sure their little plant would be putting out gasoline. Using old wood and second hand pipe, Mosher built a cooling tower. Instead of purchasing a dehydrator for $975, Mosher used a discarded boiler he found behind a laundry. Unable to find a distillation unit, which would have been $2,000, Mosher used an old gasoline truck with a 1,000 gallon tank that sat rusting in a yard in Vernon. Some of the laborers that Mosher hired were to be paid later out of earnings when the plant was on line. Meanwhile, Mosher was getting precious little sleep as he had to keep working on his farm as well as on the gasoline plant.

The optimistically named Plant No. 1 was almost finished when Mosher and Bering found out that none of the wells on the San Martinez Lease would be producing. They had a plant but nothing to feed into it. Mosher, who had never tried to get a gas contract before, decided in desperation to ask the production boss of Shell's Alamitos No. 1, a few blocks from the gas plant. Looking at the makeshift plant, the boss said he could "knock as much gasoline out of the gas with a stick" as Mosher could with his plant. He wasn't going to go for it until Mosher reminded him about the competitor who had said he would drink all the oil Shell could find, and added, "If he could be wrong, maybe you are too." The boss laughed and said that Mosher could tap into Alamitos No. 1.

This was the break they needed. By the end of May, just four months after the disastrous freeze destroyed his lemons, Mosher's gas plant was ready to test. Mosher had installed a small iron box with glass windows in the line where the newly distilled gasoline flowed. In this the first few dribbles of natural gasoline were seen, proving that the plant worked. This soon became a steady stream, and it was a good marketing tool. Many a producer on Signal Hill was convinced by this "lookbox" to sign gas contracts with Mosher.

Plant No. 1 could process only 250 gallons of natural gasoline a day. Out of this tiny start a great oil company, Signal Oil and Gas, was born. H. M. Mosher became a believer, partly by seeing the stream in the lookbox, and joined Signal's Board in 1924. For some time the company hung by a thread and was beset by controversies such as a stock fight for ownership between Mosher and Bering. In 1930 Standard cancelled its contract to buy natural gasoline from Signal, depriving the company of its largest customer. Mosher took the characteristically bold step, over the objections of many of his Board members, of acquiring a refinery and service stations, so that Signal could retail gasoline on its own. Eventually the company became stable with as many as twenty gasoline processing plants. The company branched out into exploration and production of oil. Portable drilling rigs, directional drilling, and treating sour gas were pioneered by this dynamic company. Signal ventured into other states and even countries. Signal was one of the independents behind Aminoil (American Independent Oil Company), which secured drilling rights in the neutral zone between Kuwait and Saudi Arabia in 1948. Signal purchased several other small oil companies in the 1950s and 1960s to become a fully integrated oil company. It also diversified into other industries such as shipping (American President Lines), snacks (Laura Scudders), and aviation technology (Garrett). Finally, these other enterprises became so successful that Signal ended up selling its entire petroleum operations to a Scottish Oil Company, Burmah, in 1974. Ironically, Burmah sold everything two years later to Aminoil.

Thus, Signal Oil and Gas is no longer on the roster of players in the oil and gas industry, but the legacy of Samuel Mosher and other pioneering innovators lives on. The willingness of a lemon grower to risk everything, and to completely change his life, made all the difference.

CHAPTER THREE

Above: Red Crown gasoline being put in the Spirit of St. Louis for Lindberg's historic flight, one 5 gallon can at a time. The plane carried 500 gallons. Lindberg said he used Red Crown because it would give him more flying range.

COURTESY OF CHEVRON USA.

Below: Airplane fueling with a hose.

COURTESY OF CHEVRON USA.

Others sold over 100 percent interest in their wells. Gushers, fires and other mishaps were rather common in the early, rather chaotic days of this field.

The Dominguez Field, discovered in 1923, was developed in a much more orderly fashion than Signal Hill, primarily because all of the acreage was controlled by just three companies, Union, Shell, and Associated. Wells were drilled on a 600 foot spacing rather than haphazardly on small lots. Injection of produced gas to maintain the reservoir pressure was commenced within two years, and by 1926 virtually all of the gas being produced was either reinjected or utilized. In contrast, out of 290 million cubic feet of gas produced daily at Signal Hill, all but about 70 million was released into the atmosphere.

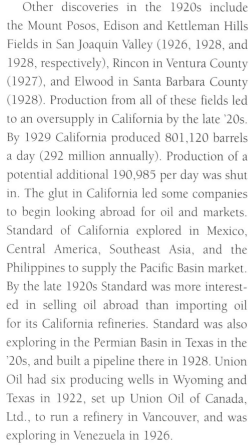

Other discoveries in the 1920s include the Mount Posos, Edison and Kettleman Hills Fields in San Joaquin Valley (1926, 1928, and 1928, respectively), Rincon in Ventura County (1927), and Elwood in Santa Barbara County (1928). Production from all of these fields led to an oversupply in California by the late '20s. By 1929 California produced 801,120 barrels a day (292 million annually). Production of a potential additional 190,985 per day was shut in. The glut in California led some companies to begin looking abroad for oil and markets. Standard of California explored in Mexico, Central America, Southeast Asia, and the Philippines to supply the Pacific Basin market. By the late 1920s Standard was more interested in selling oil abroad than importing oil for its California refineries. Standard was also exploring in the Permian Basin in Texas in the '20s, and built a pipeline there in 1928. Union Oil had six producing wells in Wyoming and Texas in 1922, set up Union Oil of Canada, Ltd., to run a refinery in Vancouver, and was exploring in Venezuela in 1926.

The 1920s saw important advances in refining and in the products offered to consumers. Oil companies started adding tetraethyl lead to gasoline to reduce engine knocking. High-octane aviation gasoline was promoted through air races, especially after the Lindberg flight in 1927 popularized aviation. Standard was quick to publicize that they had provided Red Crown gasoline for the *Spirit of Saint Louis*. Union promoted Aristo Motor Oil, the "aristocrat of motor

oils," which they said was free of asphalt. Standard began marketing an asphalt emulsion for roads and roofing in 1925.

In 1929 the first test of electrical well logging in the U.S. was done for Shell in Kern County. Electrical logging was invented by the French brothers Conrad and Marcel Schlumberger, whose company ran this first U.S. test. Electrical logging utilizes a sonde or tool that is lowered on a cable into a well. Sensing electrical or other physical properties, electrical logging can among other things provide the depths of rock formations cut by a well. From this beginning logging has become a major method throughout the world, now being done as a routine in every well when it is drilled. Logging is used by geologists to correlate formations between adjacent wells, much as Pacific Coast Oil Company tried to do using well cores at Pico Canyon in 1889.

TOWARD A NEW WAY TO PROBE THE EARTH

By 1930 the old ways of drilling seeps were in the distant past. Searching for geologic traps led to giant discoveries where geologic structures such as anticlines could be observed at the surface, generally as outcrops or topography. However, there are vast flat areas with buried anticlines that cannot be seen at the surface. There are also oil-bearing strata in coastal areas that extend offshore, as at Summerland. Early geophysical methods showed some promise of providing information about geological structures beneath the surface. Gravity or magnetic measurements made at various locations gave indications of different kinds of rocks. Passing sound waves through salt domes on the Gulf Coast (seismic refraction) could provide drilling locations for oil trapped around the domes. Unfortunately all of these methods were rather specialized and worked only in certain places. A more universal method of imaging the earth's interior was needed.

In 1921 a new invention called reflection seismology, or in the oil industry simply "seismic," was tested successfully for the first time in Oklahoma. Seismic methods in those early days were primitive compared to today. Sound (seismic) waves were generated using explosive charges in shot holes about 100 feet deep drilled by small rigs. Seismographs were of an early heavy, cumbersome type. The recording device used a light source, a mirror that was moved by the incoming electric signal, and photographic paper.

About four seismographs were deployed in a line a few hundred feet long in line with the shot point. Or, two such lines were "crossed" perpendicular to each other going away from the shot. Such cross-spreads would allow determination of the dip direction of strata. Shots were spaced from several thousand feet to a mile apart depending on the geology. In the rather simple geologic situation in Oklahoma or Texas, individual layers of rock persisted over long distances, and such a bed caused similar seismic reflections from shots over a large area. All of these arrival times of similar waves could be correlated to make a map showing anticlines and other structures that might contain oil.

In 1928 and again in 1931, seismic was tried in California, but these initial results were unsatisfactory. In California beds of sandstone and other rock are often lens-like and persist only for short distances. The seismic correlation method that worked so well elsewhere mostly failed in California

Schlumberger early logging truck, about 1932. The cable goes off frame to the right, and down a well. The two men at left are operating the recording equipment.

resistant lubricants were needed quickly, and in large quantities. For example, a single armored division required 60,000 gallons of fuel a day to fight. In addition it soon became clear that the products were not just needed, but were needed in the right place. Getting the fuel and oil to the far-flung islands and atolls of the Pacific, for example, would test the mettle of both the petroleum industry and the military. The California industry was well-equipped to assist in that part of the war because it was positioned along the rim of the Pacific Theatre.

Industry resources, including oil, tankers, and other facilities, were commandeered by the military for the duration of the war. Tankers were taken over and reassigned to the companies that owned them. A supply chain was set up in which the big, slow company tankers filled the smaller, faster Navy supply ships which in turn fueled the combat vessels. Supplying island invasions was more difficult, as oil drums had to be rolled ashore off landing craft in the middle of a battle. A plant was set up by Union Oil at Pittsburg, California (near San Francisco Bay) to fill 4,800 barrels a day. A surge of activity at this plant could indicate to the enemy an upcoming invasion, so the plant was kept secret until the end of the war, by which time 862,000 barrels had been filled with 46 million gallons of fuel.

The Petroleum Administration for War was set up by the government, with industry representatives, to put the industry on a war footing. One committee was responsible for California and other Pacific states. It came up with the concept of maximum efficient rates (MER), which was defined as the production rate for a field that if continued for six months would not reduce the ultimate production of the field.

Catalytic crackers were installed in refineries to boost the production of high-test gasoline. Other products produced for the war included plastics, synthetic textiles, and chemicals for explosives and synthetic rubber. These new technologies provided the bases for new businesses after the war. When the war finally came to an end the petroleum industry was set to expand into a new era of opportunity.

NATURAL GAS IN CALIFORNIA

Most of California's oil fields produce natural gas in addition to oil. As we have seen, in the early days gas was allowed to vent into the atmosphere in places like Signal Hill, where town lot drilling and competition with neighbors meant that one's main motivation was to produce as much oil as possible before someone else got it. Casinghead gasoline was sometimes extracted from natural gas, and the residue could be used as a fuel in oil field engines.

Natural gas as an industry in its own right did not really get under way in California until about 1910, and did not reach its stride until the 1930s when seismic exploration came to the fore. Prior to 1910 gas was sometimes collected from seeps or from shallow wells that had been drilled for water. In 1854 the courthouse in Stockton was provided enough gas for lighting by a water well nearby. This hydrocarbon-producing well was drilled at least five years before the famous Drake Well in Titusville, Pennsylvania. Gas and oil seeps were being found along the tributary rivers of the Sacramento River that were in the range of hills west of the valley. Unsuccessful wells were drilled near seeps in Colusa and Glenn Counties in 1865-1866. A gas seep was found next to Sutter Buttes, a volcanic plug in the middle of Sacramento Valley. A mine shaft was dug to find out where the gas came from; they thought the source would be oil or coal. Unfortunately, an explosion injured two miners and they had to stop. By 1890 many water wells from Tehama County south to the lower part of the Sacramento River Basin were supplying gas as well as water to farmhouses. In 1890-1891 companies were drilling wells in Stockton and Sacramento. Some water/gas wells in Stockton reached depths of 1,000 to 2,000 feet.

In 1909, large amounts of dry gas were found in the Buena Vista Oil Field, which was in an anticline parallel to and northeast of that of the Midway-Sunset Oil Field in San Joaquin Valley. By 1913 pipelines were supplying this gas to Bakersfield and

Los Angeles. Gas was discovered in the Elk Hills Oil Field in 1919. At Buttonwillow the first large accumulation of gas that was not associated with oil was found in 1927. More gas found in the Kettleman Hills Oil Field brought about the construction of a pipeline to San Francisco. By this time an oversupply of gas caused many producers to flare or vent their gas. This was a factor in the legislature's passing of the Gas Act in 1929, which required producers to utilize their gas or inject it back into the formation. Reinjection, as in the Dominguez Oil Field, helped to maintain formation pressure, making it possible to recover a greater percentage of a field's oil reserves.

A dry gas field was discovered in 1934 at Trico, about forty miles north of Bakersfield, on the basis of gas in water wells and a slight topographic rise. This field is in the central, almost flat part of the valley. Using the new tool of seismic, and adapting it to California's conditions, explorationists began finding fields in new places. In 1933 seismic helped discover gas near Sutter Buttes in Sacramento Valley. The Tracy Field was found, followed by the Rio Vista Field, the largest gas field in California, in 1936. Smaller gas fields were discovered in the northern part of San Joaquin Valley. The Santa Barbara—Ventura area is also home to gas fields, as is Santa Maria Basin. Gas in Los Angeles Basin is associated with oil.

Sacramento Valley is almost exclusively a gas province, with about eighty gas fields and only two small oil fields. About 10 trillion cubic feet (TCF) of gas, and just a few million barrels of oil, have been produced from a fairway about 200 miles long and 45 miles wide. The main reason for the lack of oil is that the Monterey Formation is confined to central and southern California. The habitat of the gas in Sacramento Valley is quite different from that of the oil in the southern part of the state. The reservoir and source rocks are much older, mainly Upper Cretaceous through Eocene (38 million years old or older). Traps were formed by folding and faulting in Oligocene time (38 to 25 million years), before deposition of the Monterey Formation to the south even began.

A formation called the Great Valley Sequence outcrops in the hills west of the valley. Consisting of interbedded sandstone and shale, the Great Valley Sequence houses both the source rocks and reservoir rocks. The strata dip almost vertically west of the valley, and "bottom out" deep under the middle of the valley, where they are buried under thousands of feet of younger rocks. The seeps are found in the hills where the Great Valley rocks are at the surface. In the valley the deeply buried rocks have no effect on surface topography, and gas traps must be located with seismic methods.

Out of 80 gas fields, only 6 have production of more than 250 billion cubic feet (BCF). The Rio Vista Field is by far the largest. By 2010 it had produced more than 3.6 TCF, three times that of the next largest field. Covering 29,000 acres in two counties, the field is located in the Sacramento Delta region about twenty miles northwest of Stockton. The confluence of the Sacramento and San Joaquin Rivers is nearby. The Town of Rio Vista is completely surrounded by the field. The field also underlies several islands of the Sacramento River Delta. Much of the area is below sea level and is protected by levees. Production significantly increased during World War II to meet demand as gas could no longer be shipped to San Francisco from Kettleman Hills. The Kettleman pipeline had been converted from gas to oil for the war effort. Rio Vista was unitized in 1965, and in 1999, Amerada Hess sold the field.

Many of the Sacramento Valley gas deposits are elusive. This is because anticlines and other traps in the reservoir rocks do not project upward into the overlying strata, at least in the central part of the valley. The resulting lack of surface expression has made Sacramento Valley a laboratory for development of better methods of imaging the subsurface, especially seismic. Seismic methods such as bright spots and AVO (amplitude variation with offset) were developed in the 1970s and 1980s using in part data from Sacramento Valley to prove the method. Today 3D seismic is used extensively to explore for new deposits in the valley.

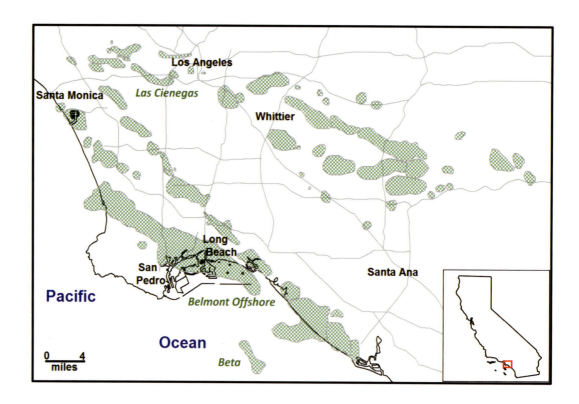

CHAPTER FOUR

PEACE, PROSPERITY AND CHANGE

OLD OIL, NEW IDEAS

Above: Three major Los Angeles Basin Fields discovered since World War II: Las Cienegas, Belmont Offshore, and Beta.

Opposite, top: Fire hydrant in Wilmington Field area. Subsidence has taken the land surface, where the dog is sitting, well below sealevel. The bow, gun turret and superstructure of a navy ship are visible behind the levy in the background.

COURTESY OF HISTORICAL SOCIETY OF LONG BEACH.

Opposite, bottom: Well heads that were raised during a landfill operation carried out to compensate for subsidence. The land surface was subject to inundation before the landfill.

COURTESY OF HISTORICAL SOCIETY OF LONG BEACH.

With the close of World War II the military demand for fuel and other petroleum products decreased to a fraction of its former self. This was more than made up for by the baby boom and new prosperity of the 1950s. New well notices in California climbed to 2,772 in 1948. Production was opened in Cuyama Valley in 1948-1949 and in Salinas Valley in the '50s. The natural gas province in Sacramento Valley was greatly expanded in the '50s and '60s. California's gas production, after slowly declining through most of the 1950s, went from about 450 billion cubic feet (BCF) annually in 1958 to 725 BCF ten years later. The Paloma Field in San Joaquin Valley was the scene of a new world drilling record, 20,521 feet.

On the other hand, mature areas saw a tapering off. Los Angeles Basin, home of billion barrel giants, has seen only three substantial discoveries since the war, Belmont Offshore in 1948 (75 million barrels, actually an extension of the Wilmington Field), Las Cienegas in 1960 (75 million), and Beta offshore in 1976 (220 million). Total new field discoveries in the basin in the last seventy years amount to only about half a billion barrels, a small fraction of the 10 billion or more barrels that have been produced.

Far more has been added to California's reserves since the war by developing new enhanced recovery methods than by exploring for new fields. U.S. Department of Energy data indicate that only 3 percent of reserves additions in San Joaquin Valley in the 1980s came from new field discoveries. The rest was from enhanced recovery and development of new pools in existing fields. Giant fields discovered in the early days, such as Kern River, Coalinga, Midway-Sunset, South Belridge, and Elk Hills saw the biggest gains. For example, Midway Sunset went from an estimated ultimate recovery (EUR) of 1 billion barrels in 1950 to 3.5 billion in 2000, and

Kern River went from 500 million to 2 billion in the same period. Similar increases occurred at Huntington Beach, Ventura, and San Ardo in Salinas Valley. Wilmington has also benefited, although its major gains have been from field extensions to the southeast—Long Beach Harbor (THUMS) and Belmont Offshore.

Sometimes producers got enhanced oil recovery as an additional benefit from oil field operations that were intended for another purpose. The Wilmington Field had been producing at top speed for the war effort. After the war strange events were occurring at the Naval Shipyard and other facilities in the oil field area. Railroad tracks buckled. Sewer lines broke. Walls cracked. Surveyors found that the land was subsiding as much as two feet per year in some places by the early 1950s. Dikes were built to keep the ocean away from the shipyard as well as from factories and warehouses employing 1,000 people. Horizontal movement associated with the subsidence caused a large amount of the damage. The oil field itself was affected. Well casings were crushed or sheared off, and many producing wells had to be abandoned.

WELL RAISING, AREA 19A
LONG BEACH HARBOR
MARCH 9, 1954

By this time it was known that more reserves existed to the southeast under Downtown Long Beach and the adjacent harbor. If the subsidence problem could not be solved perhaps this rich resource would never be developed.

Richfield and other operators realized that production of oil from the field's shallow, thick, unconsolidated sands was the cause of the problem. Reduction of formation pressure as the oil was produced allowed the sands to compact. In 1955 the president of Richfield, Charles Jones, proposed a solution in a speech at the Long Beach Rotary Club. He suggested that water be injected into the reservoir to increase the pressure and arrest the subsidence. Because there were a hundred operators and over 1,000 property owners, legislation was needed to unitize the field. Agreements were made on how the City of Long Beach and the operators would work together to make the biggest waterflooding project in the world work.

Finally, water injection got underway in October 1959. Not only did it arrest the subsidence of the land, the water also pushed oil ahead of it to producing wells, leading to the recovery of an another 500 to 700 million barrels. In addition the City of Long Beach opened bidding on the southeastern extension of the field when it was clear that the subsidence had stopped. This added another 1 to 1.2 billion barrels to the total reserves of the field.

Other ways of getting oil out of the ground were tried in the California oil fields, leading to enhanced oil recovery methods now used around the world. By 1950, as much as 40 billion barrels of heavy, sulfur-rich crude had been discovered in California, but by some estimates only 10 percent of it had any chance of being recovered. One approach was to heat up the oil to reduce its viscosity. Several ways of doing this were experimented on in the laboratory, then tried in small pilot projects in oil fields.

One idea was to set the oil sand on fire. Called "fireflooding," this was done by pumping air into the formation. Oxidation of the oil heated the formation enough to cause spontaneous ignition. It was hoped that about

Well on a long, narrow (about 100 feet wide) lease that included the tracks of the Southern Pacific Railroad. The original lease holder was Jack Herley Operations. Operations coexisted at close quarters for decades with both railroading and agriculture.

10 percent of the oil would burn (primarily the tarry, less desirable fraction) and heat the rest so that it would flow more easily to the producing wells. In 1956 General Petroleum ran a test in the South Belridge Field using an injection well in the center of a 330 foot square with a producing well on each corner. Another test was done in the same year by Standard of California at Midway-Sunset with an injection well and three producers in a 200 foot circle around the injector. Increases in production were observed but the method was expensive. It never really took off, especially with smaller operators.

Another method was bottom hole heating, in which a fluid, either oil or water, was heated at the surface and pumped via tubes inside the casing down to a heat exchanger unit at the bottom of the hole. The surrounding formation was heated by conduction. This was much cheaper than fireflooding as it used equipment that could be moved from well to well. New injection wells did not have to be drilled. The problem was that it could heat the formation only up to a short distance from the borehole. A better way of heating large volumes of oil sand was needed.

That idea was steamflooding. It was tried first by Shell in 1960 in the Yorba Linda Field. This field, in the eastern part of Los Angeles Basin, was discovered in 1930 with reserves of about 90 million barrels of 12 gravity oil. Shell's pilot project, estimated to cost $150,000, involved pumping steam into the shallow (600 foot) pay zone to heat the heavy oil. Shell kept the project secret, and began another pilot on the 15 gravity oil at Coalinga later in the same year. This test called for an injection well in the 1,000 foot deep pay zone, four observation wells 90 feet away, and two producers on opposite sides 180 feet from the injector. Shell kept mum on this test too, but the continuing steaming operations at both sites could be observed by competitors.

Meanwhile, Tidewater was experimenting with hot water flooding at Kern River. This method had a problem with channeling, in which the water created pathways that left much of the surrounding sand unaffected. Tidewater began experiments with steamflooding.

Although Shell and Tidewater did these tests in secret, others eventually figured out what they were doing. What soon became obvious to everyone was that a large number of leases were suddenly being sold, and that the buyer was usually Shell or Tidewater. Many leases were going for prices far above the value of current production from stripper wells. Clearly something was up. Other companies started buying leases. Steamflooding was much cheaper than fireflooding, and even a relatively small operator could do it. People talked about ten-fold increases in production of stripper wells at Midway-Sunset and South Belridge. By 1964 the secrecy had melted away. Publications like *California Oil World* were talking about steamflooding.

The habitat of the heavy California oil makes it especially suited for steamflooding. For example, at Midway-Sunset most of the pay sands are at 7,000 feet or less, and reservoir temperatures are relatively low, below 100 degrees F. Porosity and permeability are very good. Under these conditions the steam can substantially increase the temperature of the cool, heavy crude. Waterflooding and fireflooding had been tried at Midway-Sunset as early as 1954 and 1960, respectively. Steamflooding, which started in 1963, was expanded to five different pools by 1970. A rapid increase in EUR of about 500 million barrels occurred in 1969-1970. Later stepwise increases were due to discovery of new pools and further enhanced recovery.

PEOPLE-FRIENDLY DRILLING

In the early town-lot days wells were drilled with little regard for the effects of noise, fumes, runaway spills, fires, gushers, and other events that at the time were considered normal oil field occurrences. Derricks seemingly appeared almost anywhere in Los Angeles, often close to residences, parks or businesses. As the city grew it became apparent that oil field operations would have to adapt to an urban environment. This led to

Steam injection well, Kern County, in the 1960s.

PHOTO FROM BILL RINTOUL COURTESY OF GENERAL PRODUCTION SERVICES.

innovations in California that were eventually adopted in many other places in the world.

One of the first times a well was drilled with special consideration for surrounding people and businesses was right in the middle of World War II. In 1943 at First and Gardner Streets, fifty years and three miles removed from the boisterous Los Angeles City Field, Shell drilled the Verne Community No. 1 Well in the old Salt Lake Field. This deep test was appropriately named after the author of *20,000 Leagues Under the Sea*. Shell used electric motors enclosed in a noise and fume proof covering. The drilling site was also surrounded by a high fence. Although the 7,924 foot test was a dry hole, it introduced the idea of beautifying and sound-proofing an urban drilling site.

In 1949 Union Oil began developing the Sansinena Field near La Habra, which had been discovered in the closing days of World War II. Mineral rights for this 3,400 acre tract had been purchased by Lyman Stewart back

Opposite, top: Bolsa Chica section of the Huntington Beach Field, in close proximity to a housing development. Also visible is one of the offshore platforms, and behind it Santa Catalina Island.
PHOTO BY AUTHOR.

Opposite, bottom left: Building to house derrick at Occidental's oil island. Installed in 1966, the site is now operated by Pacific Coast Energy Company. Complete with a vine-covered wall, shrubbery and trees, the site blends into the otherwise typical west Los Angeles street scene. The only sound emitted is a faint hum of machinery. Interestingly, the three streets that converge here are all named after oil pioneers.
PHOTO BY AUTHOR.

Opposite, bottom right: A covered derrick.
PHOTO FROM BILL RINTOUL COURTESY OF GENERAL PRODUCTION SERVICES.

Above: Oil in Surf City. Covered rig along Pacific Coast Highway, Huntington Beach.
PHOTO BY AUTHOR.

Left and below, three views of the same building: Oil island at Pico Boulevard and Genesee Avenue built by Standard Oil. Now operated by Freeport-McMoRan, it can house two drilling rigs at a time. The building appears to have a front entrance like an ordinary office building. Landscaping is meticulously maintained.
PHOTOGRAPHS BY AUTHOR.

in 1902 without telling his board of directors. Union had done little with the property even though it was on-trend between the Brea-Olinda and Whittier Fields. Years later the surface rights were sold off and affluent residential areas developed. The homeowners were convinced to allow drilling only when Union offered to give them royalties (even though it owned the mineral rights) and pledged to drill directionally from "islands." Using sites hidden in canyons, Union developed the field into a 11,000 barrel per day producer by 1957. Derricks were covered with glass cloth and fiberglass sheeting that was painted green on the outside to blend in with the surroundings. Exhausts were equipped with mufflers. Instead of using mud pits, drilling fluids were stored in tanks and waste was hauled away in tank trucks.

This successful development provided an example for further drilling in the more densely populated Los Angeles. In 1950 the Los Angeles City Council adopted rules permitting drilling only from "islands." Electric motors, soundproofed rigs, disposal of waste, and other practices used at Sansinena were required. Some interesting

Right: Ring of boulders will form the perimeter of a THUMS island. The dredge is beginning to fill the interior of the island. Downtown Long Beach is in the background.

Below: Map of part of west Los Angeles showing tracklines of wells drilled from Occidental's oil island. Major streets are shown in orange. Other oil islands are nearby, and together they provide coverage of the Beverly Hills Field. Wells to the southwest were drilled from the Hillcrest Country Club, and wells along the east margin of the map were drilled from Standard's oil island farther east at Pico Boulevard and Genesee Avenue.

drilling sites were used, such as the Twentieth Century Fox Movie lot, which was in the old Beverley Hills Field. This field, discovered in 1900, was fading in the 1950s when the Universal Consolidated Oil Company drilled 5,000 feet deeper than the original 2,500 pay zone. They brought in a 525 barrel per day producer. Fifty-two wells were drilled from two islands on the studio lot. Concrete retaining walls and other soundproofing were essential so as not to disturb nearby movie-making operations.

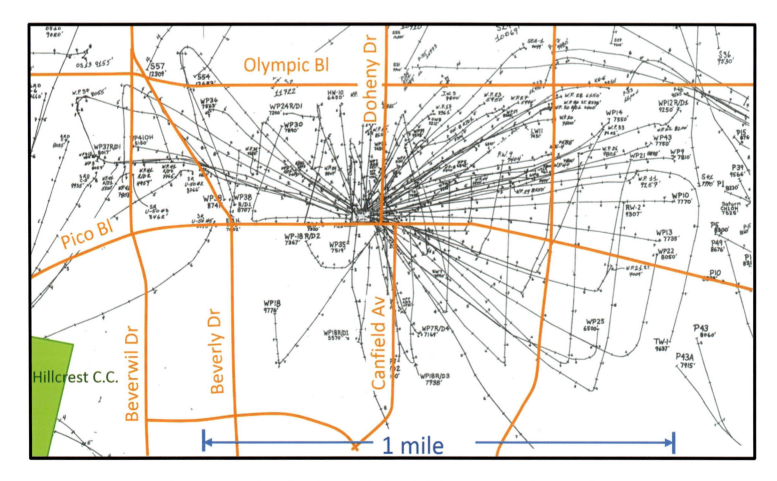

Another drilling site was the Hillcrest Country Club, whose membership list included show business names and high finance people. Signal Oil and Gas wanted to drill directionally at a site 100 yards from the clubhouse. The country club managers agreed to the deal as a means of covering rising expenses. Jack Benny, a member of the club, found a way as usual to make a joke of the situation, saying, "Perhaps if we sign up with Signal, we will be as rich as Bob Hope and Bing Crosby someday."

Drilling islands sprouted all over the city. Three such sites were built along Pico Boulevard, appropriately named after the early oil pioneer. Two were made to look like multi-story office buildings. Occidental Petroleum, which built one at a cost of $1,000,000, called it the "world's first architecturally designed oil derrick." Another site, owned by Standard of California, was so well disguised that a group of visiting executives from another company reportedly drove by it three times before a policeman convinced them that the "office building" was really an oil island. The Las Cienegas Field was developed entirely from oil islands, including one of the three on Pico Boulevard. Yet another oil rig, built by Mobil at Venice Beach, looked like a lighthouse.

One of the best known, and most elaborately disguised, oil operations is the set of four oil islands operated by THUMS in Long Beach Harbor. These are islands in the real sense of the word, artificially made to develop the southeast extension of the Wilmington Field. When it was clear in 1964 that subsidence in the Wilmington Field was under control, the City of Long Beach and the state, co-owners of the mineral rights, opened the extension to bidding. Five companies joined together to submit the winning bid, which had a razor-thin profit margin. The companies were willing to take a small percentage because oil was already known to exist in the area where the Wilmington Anticline continues to the southeast, and because transportation and refining facilities existed close at hand. The five companies in the original consortium were Texaco, Humble, Union, Mobil, and Shell; hence the acronym THUMS.

THUMS constructed four islands in Long Beach Harbor by making rings of boulders quarried on Santa Catalina Island, then filling

Piers. The first "platform" was a portion of a pier like these.
COURTESY OF CALIFORNIA OIL MUSEUM, SANTA PAULA, CALIFORNIA.

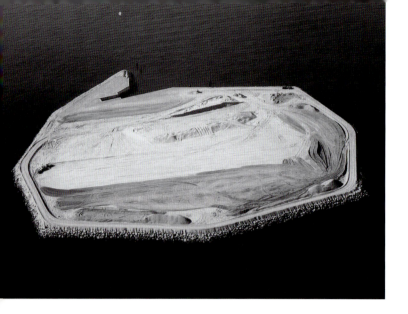

Top, left: THUMS island newly filled with dredged material.

COURTESY OF HISTORICAL SOCIETY OF LONG BEACH.

Top, right: Derricks and other facilities are being installed. Palm trees that will make up part of the landscaping have been planted. The derricks are covered to varying degrees with structures to make them look like condominiums.

COURTESY OF HISTORICAL SOCIETY OF LONG BEACH.

Middle: Nearly completed THUMS island as it appears from ground level. More work is needed to completely extend the cosmetic ring of trees and shrubs around the island.

COURTESY OF HISTORICAL SOCIETY OF LONG BEACH.

Bottom: THUMS Island Chaffey. The four islands have been named after U.S. astronauts who gave their lives in the early part of the NASA space program. Tracks are visible on which the derrick can be moved to access well sites. Tanks and other supporting equipment are in the interior of the island. Landscaping and cosmetic structures can be seen on the perimeter.

COURTESY OF BRUCE PERRY,
CALIFORNIA STATE UNIVERSITY LONG BEACH.

the interiors with sand dredged from the harbor floor. The derricks were disguised with painted balconies and lighting to look like luxury condominiums. Only a few derricks were needed on each island as they could be moved to any desired location on a set of rails. Directional wells were drilled from surface locations only six feet apart. The perimeter of each island is decorated with palm trees, oleanders and other shrubs, and there are waterfalls illuminated at night. Although many of the locals know that the towers are derricks, they seem to be pleased with the esthetics, as are the many visitors and tourists. By the end of 1967 THUMS was completing about one well a day, some of them bottoming out under the streets of downtown Long Beach. Production was 50,000 barrels a day and rising.

THE MARCH OFFSHORE

As has already been mentioned, the U.S. offshore petroleum industry started in 1897 when the first of many wooden piers were built to support small derricks drilling shallow wells off of Summerland. Seeps seen at low tide indicated that the producing beds on land continued offshore. These were small wells producing an average of 1 barrel a day. They were profitable because many of them could be operated cheaply with jacklines and there was a local market for the oil. Although storms destroyed many of the wells, production continued until the last well was lost in 1939.

After a tideland leasing act was passed by the legislature in 1921, drilling was done at Rincon and Elwood, near Ventura and Santa Barbara respectively. Steel and concrete were used in the piers, which went as far as 2,300 feet into the ocean. In 1932 the first platform not connected to shore was built at Rincon, actually a section of a pier. This structure survived until 1940, when it was brought down by a winter storm. All of this activity was extensions of known onshore fields. The first discovery of an entirely offshore field in California was Summerland Offshore in 1957.

A scene of much early offshore development, and controversy, was Huntington Beach. The earliest offshore production was apparently known only to the operators and not to the industry generally. Two wells drilled in the early '30s in the townlot section had unusually high production for townlot wells. The Superior Babbitt No. 1 at one point reached 1,450 barrels a day, and the Wilshire H. B. No. 15 came in at 4,815 barrels a day. It was found that this well had drifted 1,400 feet offshore. It was suspected that other operators had trespassed into offshore state lands, so the state got an injunction to force one of them to do a directional survey. An offshore pool in an anticline extending obliquely into the ocean from the main Huntington Beach structure had been tapped. The Babbitt Well was determined after the fact to be the discovery well. In 1934 the production rate in this pool was over 20,000 barrels a day, far greater than any previous offshore California production. Eventually the state and operators reached a compromise in which the state received royalties on the oil. A state lands act passed in 1938 provided a more orderly way of leasing and drilling offshore, setting the stage for great expansion in the post-war era.

In the late 1940s, Marine Exploration Company (later Monterey Oil Company) had a state tidelands lease off Seal Beach. Seal Beach had an ordinance against drilling in the city limits, so Marine Exploration had to drill a well from an onshore site in adjacent Long Beach. This remarkable well, drilled in 1948, was spudded one and a half miles inland and whipstocked 9,271 feet, almost two miles, toward the offshore lease. The length of the borehole was 12,180 feet and the true vertical depth was 5,700 feet. They hit an oil sand and the well came in at 30 barrels a day, not a barn-burner, but encouraging.

The City of Seal Beach claimed jurisdiction out to three miles offshore, a claim that Marine Exploration decided to test. They built an island one and a half miles offshore. The city filed a criminal complaint that

Piers at Elwood, 1934.
COURTESY OF CALIFORNIA OIL MUSEUM, SANTA PAULA, CALIFORNIA.

Ever since the 1969 offshore spill, one of the most politically and emotionally charged issues of our time has been how to balance the need to produce and import energy with having a clean, safe coastline.

Major sources of oil and natural gas in offshore marine water are naturally occurring seeps. Lines of seeps are found in Santa Monica Bay and especially in Santa Barbara Channel. Isolated seeps have been found in San Pedro Bay and around the Channel Islands. Lines of seeps occur along the axes of two parallel anticlines up to two miles off Coal Oil Point west of Santa Barbara. Globules of tar are commonly found on nearly any beach in southern California, and almost any resident who goes to the beach will step on it sooner or later. At Carpinteria State Beach, oil oozes from active seeps in the beach bluffs, and great mounds of immobile tar, looking much like lava flows, extend into the surf zone.

Platform Holly, in state waters off Coal Oil Point, west of Santa Barbara. The platform is now operated by Veneco.
COURTESY OF VENECO.

Some of the offshore seeps are quite active, especially off Coal Oil Point, discharging large amounts of natural gas and smaller amounts of oil. In 1982 ARCO installed two seep tents, steel pyramids 20,400 square feet in area, on a seep about a mile southeast of Platform Holly. These tents initially collected about 1 MCF of gas per day. The rate doubled before beginning a long decline

in 1989. Platform Holly is located on another seep. Both platform and tents are right on the axis of the South Elwood Anticline.

The seeps can be mapped using sound waves in the ocean. Echo sounders detect sound waves that reflect off of the column of bubbles of gas in the water above a seep. Running a small boat in a grid pattern allows one to build up a map. Comparison of mapping surveys conducted from 1973 to 1995 showed that seepage within two-thirds of a mile from Platform Holly had almost disappeared by 1995. In contrast, other seeps along the south Elwood Anticline remained about the same. Between 1967 (when Holly began producing) and when the last seep survey was done in 1995, 50 million barrels of oil, an equal volume of water, and 30 BCF of gas were produced. Reservoir pressure decreased about 35 percent in the same period. It seems likely that production of fluids from the reservoir caused a reduction in seepage.

Sometimes it is possible to use chemical analysis to determine whether oil pollution on a beach comes from a spill or from natural seeps. On February 7, 1990, the *American Trader* ran aground about two and a half miles off Huntington Beach, spilling 9,381 barrels of Alaskan crude. Much of this oil washed ashore in a storm 7 and 8 days after the accident. The last of the oil came ashore on day 12. Skimming offshore removed about 38 percent of the spilled oil before it reached shore or sank to the bottom. Another 9 percent was removed by cleanup operations onshore. Chemical analysis of biological marker compounds, leftovers of original organic molecules that were the raw material that became oil, serves to "fingerprint" or identify the oil. The Alaska oil has a fingerprint quite distinct from that of the local oil. This made it possible to show that the spilled oil had disappeared from the beach by day 20 after the spill. Tar globules found after that were "background" seep oil that is virtually always present.

These examples show how modern technology—chemical analysis, marine acoustical surveying, and other methods—can help us better understand how normal oil field operations, accidental spills, and natural seepage all affect coastal pollution. This could support rational planning for offshore production and coastal terminals.

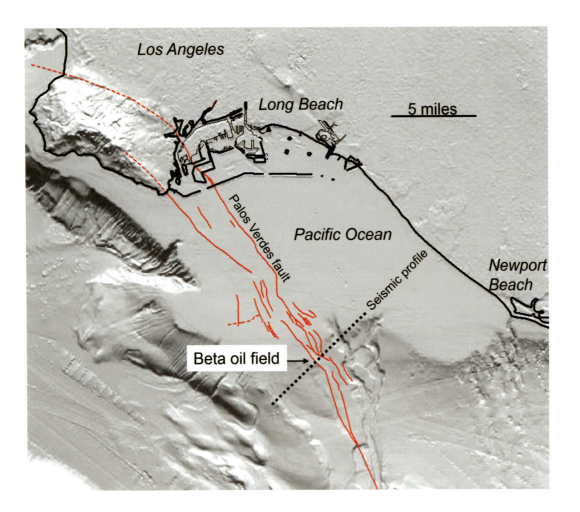

Los Angeles

Long Beach

5 miles

Palos Verdes fault

Pacific Ocean

Seismic profile

Newport Beach

Beta oil field

Left: Map showing track of seismic reflection profile shot across the Palos Verdes Fault.
BATHYMETRIC IMAGE FROM U.S. GEOLOGICAL SURVEY.

Below: Seismic profile showing prospective oil traps formed in part by the Palos Verdes Fault. This was developed as the Beta Field.
SEISMIC DATA FROM NATIONAL ARCHIVE OF MARINE SEISMIC SURVEYS, (NAMSS). GEOLOGIC INTERPRETATION BY AUTHOR.

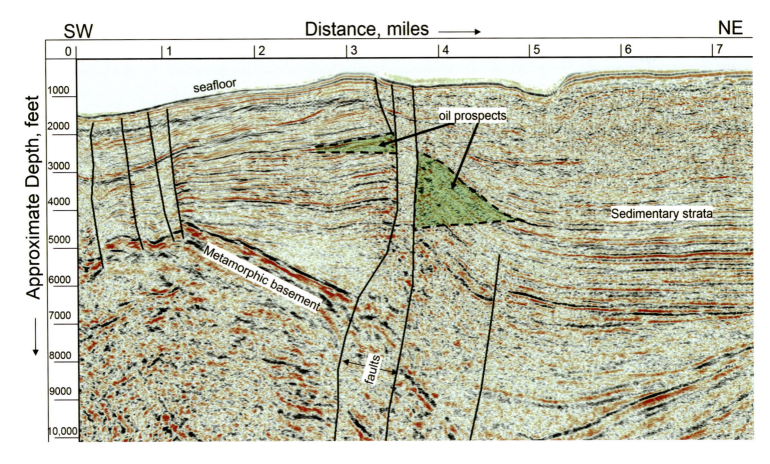

SW   Distance, miles ⟶   NE

0   1   2   3   4   5   6   7

seafloor

oil prospects

Sedimentary strata

Metamorphic basement

faults

Approximate Depth, feet

1000
2000
3000
4000
5000
6000
7000
8000
9000
10,000

Gas line at a Chevron station in Los Angeles during the 1973 oil embargo.

COURTESY OF CHEVRON USA.

federal government to hold a lease sale for a drainage tract in 1966. In another lease sale a little over a year later, it offered 110 tracts of outer continental shelf (OCS) land in Santa Barbara Channel. This sale brought in the highest bid for any offshore sale up to that time. Interest in offshore California was at a peak. By the late 1970s companies were drilling test holes throughout the OCS. Thousands of miles of multi-channel seismic reflection profiles were acquired to map possible oil traps. Some wells and seismic were done as far offshore as Cortes and Tanner Banks, 100 miles from the coast and 45 miles from San Clemente Island, the nearest land. This remote (for that time) area of violent seas, high risk, and enormous logistical challenge would probably have required a supergiant discovery larger than Wilmington to justify development.

Explorationists made significant discoveries in the OCS in the 1970s, including the Beta Field on the San Pedro Shelf. However the pace of development of platforms and other facilities slowed before picking up again in the 1980s. Four platforms were installed in the 1970s, compared to eighteen in the 1960s (including four islands) and fourteen in the 1980s. The 1970s slowdown was in part due to the 1969 offshore oil spill in the OCS Dos Quadras Field near Santa Barbara. Although the flow of oil was under control in about eleven days, heavy seas moved the oil slick onto the shore, where it was sprayed onto sea cliffs and homes by wind and waves. Struggling wildlife and

coastal damage received intense media attention, and pockets of strong opposition to the petroleum industry in general developed. Litigation by environmental groups delayed until 1977 the installation of the third platform in the Dos Quadras Field. However, the offshore industry recovered, and most platforms off the California Coast were permitted and built after the 1969 spill with the most recent platform erected in 1989. Furthermore, drilling from existing leases both in state and federal waters is allowed today.

A NEW WAY OF REFINING

Refining the heavy, California oil had always been a problem. Simple distillation would separate out gasoline or kerosene range compounds, but left a large amount of heavy fuel oil and a smaller amount of mostly useless tar. By the 1950s refineries at Union Oil were able to produce about 55 barrels of fuel from every 100 barrels of crude. This mix did not satisfy the demand, which was continuing to shift to gasoline and jet fuel. A method called "hydrocracking" could break down the large molecules of fuel oil and tar to smaller molecules that make up gasoline and jet fuel. The problem was that hydrocracking had to be done at extremely high pressures, up to 10,000 pounds per square inch. The large vessels needed for a commercial process could not take this kind of pressure. Some other solution was needed. Union's researchers found a catalyst that would promote the

reaction at much lower pressures. They named the new process "Unicracking." It could actually produce 115 barrels of gasoline from 100 barrels of crude, without leaving a tarry residue. This "magic" resulted from the fact that the gasoline takes up more volume than the original oil. At long last the old problem of refining California's oil was solved!

There was a problem however. The researchers found, after a full size containment unit for a cracking unit had been ordered, that the catalyst breaks down into a fine powder. They had to spend six months of non-stop work to find a bonding agent that would hold the catalyst together. They finally found it in aluminum oxide, a very simple, common compound.

It turned out that Standard of New Jersey was working on the same cracking problem. The two companies decided to pool their patents and together license the process to other companies. Soon after Unicracking came on line in 1964, ten other companies were building cracking units.

THE TURBULENT 1970s

The 1970s were a difficult time for the petroleum industry in California and elsewhere. In 1971 President Nixon announced a wage and price freeze. Although general wage and price controls were eliminated in 1973, price controls were kept on crude oil until 1981. The Arab oil embargo in 1973 caused extreme shortages of refined products, and the well-known gas lines. In March 1974 the embargo ended, allowing plentiful supplies of fuel to return. Price controls on domestic oil discouraged exploration in the U.S., including California. Imports of foreign oil increased dramatically in order to satisfy demand. Another shortage, with gas lines in some places, occurred at the end of the decade partly as a result of the revolution in Iran. Finally, in 1979 phased decontrol of prices began, although a "windfall profits" tax was proposed to compensate for the increased income oil companies would receive from the higher oil prices. Ultimately, price controls were completely and immediately lifted when President Reagan took office in January 1981.

It would seem that the 1970s was an experiment in detailed government management and control of the oil industry. It did not work out very well. The main things it did were to discourage domestic exploration and production and increase our consumption of imported oil. A long lasting effect of the decade was an increasing concern in the public for environmental problems. Much of this concern at the time had to do with air pollution, notably in the Los Angeles Basin. Clean Air Act amendments and the creation of the Environmental Protection Agency led to catalytic converters in vehicles and removal of lead from gasoline. This required refiners to come up with new processes to manufacture gasoline that would perform well with minimal engine wear.

California saw a decrease of about 15 percent in its production over the first half of the 1970s. Production then recovered to the previous long-term rate of increase that had prevailed for more than seventy years, ever since the discovery of the Los Angeles City Field. However, change was coming to California's petroleum industry in the next decade and beyond. Rising challenges in this now mature petroleum province would bring new players and innovations to the fore. New environmental concerns and often vocal critics required companies to focus significant effort on how their projects affected surrounding communities. In spite of the changing business climate the industry was able to continue to be a critical part of the state's economy into the 1980s and beyond.

CHAPTER FIVE

THE PATH TO THE FUTURE: 1980 TO PRESENT

In 1980 a new era was about to begin in California's petroleum industry. Many of the stalwarts such as Union, Getty, and Signal Oil had disappeared or would do so in time. New names would take the stage. Businesses in the state and in the whole world had been disrupted by embargoes, inflation, and shortages in the 1970s. The environment of government regulation and taxation was highly unpredictable, making long-range planning difficult. California's production rate reached its peak and began a long-term decline in the mid-'80s. For more than twenty years, reserve increases had been due largely to enhanced recovery methods rather than discovery of new fields. Most of the giant fields like Wilmington and Midway-Sunset were mature. Production was declining despite the best efforts of operators.

Nevertheless a bright future lie ahead. Federal regulation of the industry rationalized in the early '80s allowing new ideas and new innovators to come to the fore. California's oil fields are mature, but given California's complex geology and technological innovations, oil fields with stacked pay and previously unreached deposits have largely stabilized what would typically be declining production. California in the last several years has been the third highest producer in the U.S., after Texas and North Dakota.

California independent Venoco was founded in 1992 by Tim Marquez with $3,000 in capital. The company now has several onshore and offshore fields, and in 2006, expanded to Texas.

PHOTOGRAPHS COURTESY OF VENOCO.

UNION OIL:
AN ERA ENDS AFTER 115 YEARS

Union Oil, or Unocal, passed into history in 2005 when it was absorbed by Chevron. In its long and storied existence it survived four takeover attempts. Each crisis played a role in shaping this company. The first was the proxy battle precipitated by Thomas Bard in 1899, the second in 1922 was by a group of foreign investors known as Royal Dutch Shell, and the third was by Phillips Petroleum in 1959.

The Shell struggle received a great deal of media attention. The patriotic zeal of California's citizenry was appealed to in order to prevent the company from falling under foreign control. The Los Angeles Chamber of Commerce raised "...the real danger of foreign domination of this company, which has been heretofore 'of and for' Californians." The *Los Angeles Express* said, "...Every stockholder owes it to his pocketbook, to California and to the nation to keep the American flag flying over California's oil fields." When it came down to the stockholders' vote on March 20, 1922, Union had won by 25,000 shares. It would be Lyman Stewart's last great fight, and his final victory. He died in 1923 at the age of eighty-three. It is likely that this episode helped to cement Union's reputation as a California company. It retained this regional stamp for decades even as it began to expand into overseas operations.

The company's fourth struggle to maintain its independence would help to redefine the company and contribute to its ultimate fate. In late 1984, while engaged in a bid to take over Phillips Petroleum, T. Boone Pickens and Mesa Petroleum began buying shares of Union. Union, reorganizing into Unocal, a Delaware corporation, began going by that name during the fight with Pickens. Delaware's laws have been crafted to make it harder for an outsider to take over a company and are why many companies have incorporated there. After acquiring 13.7 percent of Unocal's stock, Mesa announced that it would buy another 37 percent at $54 a share. Unocal CEO Fred Hartley decried Pickens as a corporate raider, saying that his method was to buy stock in a target company with borrowed money, then dismember the company by using its assets to pay off the loans. Said Hartley, "Mesa was coming not to build, but to destroy. They were out to loot and liquidate Unocal."

Unocal countered Pickens' hostile tender with an offer of its own: to buy all remaining shares at $72 if Mesa succeeded in getting enough shares to control the company. This would leave Unocal with so much debt that it would not be a desirable target for Pickens. A court battle ensued in Delaware over whether Unocal had the right to deny the $72 offer to Mesa for its shares. The number of law firms and lawyers involved was phenomenal and they worked around the clock. They set up a special room at corporate headquarters in Los Angeles called the "Dungeon" where they kept documents under tight security.

In May 1985, Unocal won both the court case and the proxy fight, but it was a costly victory. Aside from a six month disruption of company affairs, Unocal's debt went from $1.2 billion to $5.3 billion. This was money that could not be spent on acquiring new leases or developing existing fields. Although they later restructured their debt, it took them at least fifteen years to pay it off.

In the 1990s Unocal went overseas in a big way, acquiring a large offshore natural gas concession in Thailand and other leases in Asia and Latin America. Domestic exploration was restricted to the Gulf of Mexico. To free up capital, Unocal disposed of all of its California fields by 1996. In 1997 the refineries, gas stations and transportation facilities were sold to Tosco, along with the "76" logo so familiar to Californians. The research center in Brea was closed. Even the corporate headquarters in downtown Los Angeles was sold and the company moved to nearby El Segundo. This reduced the company's debt to about $2.2 billion. In the early 2000s Unocal's reserves had declined, although the company was making more money because of the increased price of oil. New projects, mainly in Asia, were about to come on line, promising an increase in production.

Unocal had been radically transformed. Most of its assets were overseas and in the Gulf of Mexico. It had divested itself of virtually all of its U.S. refining and marketing assets. Its great discoveries at Orcutt, Midway-Sunset, Brea-Olinda, Santa Maria offshore and many others were now under the flags of independents. Although its headquarters was still in California, nearly all of the California "stamp" of Unocal was gone. Lyman Stewart would have been amazed, although he might have recognized the risk-taking culture, especially in the foreign projects.

With its new mix of assets and petroleum reserves Unocal was a ripe takeover candidate and Chevron saw it as a good fit. A deal was reached in 2005 for some $18 billion in stock and cash. A company partly owned by the Chinese government made a counter offer, sparking a controversy reminiscent of the one involving Royal Dutch Shell in 1922, although it was less intense. Chevron won, ending Unocal's remarkable 115 year span. In a sense it was coming full circle as Pacific Coast Oil Company, Chevron's forerunner, had given Stewart and Hardison their start in Pico Canyon back in 1883. Unocal may be gone, but its contributions to California's petroleum industry have had a permanent imprint.

*Well on Signal Hill, operated by
Signal Hill Petroleum.*

Operating in mature areas such as Los Angeles Basin was becoming less profitable for the large oil companies, especially after the drop in oil prices in 1986. These companies determined that the best strategy was to focus outside of California, and needed capital for large, higher-margin ventures in the Gulf of Mexico or overseas. The '80s ushered in an era of mergers and attempted takeovers, placing many companies in temporarily delicate financial positions. In 1984 Texaco bought Getty, a Los Angeles-based company. This controversial move resulted in a $3 billion legal judgement for Pennzoil, which had made a previous merger offer for Getty. In 1985 Chevron acquired Gulf following a takeover attempt of Gulf by T. Boone Pickens and Mesa Petroleum. Royal Dutch Shell merged with several other companies in the mid-'80s. Occidental (Oxy) merged with

Cities Service, another company that had been targeted by Pickens. Later, Chevron would merge with Texaco, and Exxon with Mobil. In 1985, Union barely survived yet another hostile takeover bid by Pickens, but it was weakened in the process. Eventually Union would merge with Chevron. All of these companies had mature properties in California, and many of these holdings could be used as a source for much needed cash if they were sold.

When a need arises, the business world finds a way to fill it. Independent oil companies arose to take over many of the California fields from the majors. These independents could focus on new ideas to breathe new life into the old fields. There are several breeds of independents in California. Some were formed from scratch specifically in order to revitalize old fields. In the early '80s, Hal Washburn and Randy Breitenbach were roommates at Stanford studying petroleum engineering. At the time the oil industry was booming, and the two students chose the energy sector over the up-and-coming Silicon Valley tech sector in Stanford's back yard. By the time the pair left Stanford the oil patch had hit a downturn. Despite the down times, Washburn and Breitenbach saw an opportunity in California fields, which still had a lot of oil that could be recovered through advanced technology even though they had been producing for decades. They formed Breitburn Energy in 1988, later acquiring the West Pico and Sawtelle Oil Fields from Oxy. Within a few years they had proved the value of California's oil fields by quadrupling the estimated reserves of those two fields. In 1999 they bought Santa Fe Springs and Rosecrans from Texaco. Today Breitburn operates 572 wells in California, producing 5,500 barrels a day. It has also leveraged the experience gleaned from California's complex geology and expanded its oil operations to Colorado, Texas, Oklahoma, Michigan, Illinois, Florida, and elsewhere.

Other independents were formed out of a major for the purpose of taking over its California operations. In 2014 Oxy was one of the largest oil and gas producers in California, with thousands of wells in five giant fields

Left: Arroyo Grande Field.

Below: Wells on the Maino Lease, Arroyo Grande Field, near Pismo Beach, Santa Maria Basin. Formerly operated by Grace Petroleum, the field is now operated by Freeport-McMoRan, which is using steamflooding.

in the Los Angeles Basin, San Joaquin Valley, Ventura, and Sacramento Valley. Oxy had taken over THUMS Oil Company and was thus operating the Long Beach unit of the Wilmington Field. In November 2014, Oxy formed California Resources Corporation (CRC) out of its California holdings, giving Oxy shareholders 0.4 shares of CRC stock for every Oxy share they owned.Operating exclusively in California, this new company has 19,800 gross drilling locations on 2.3 million acres.

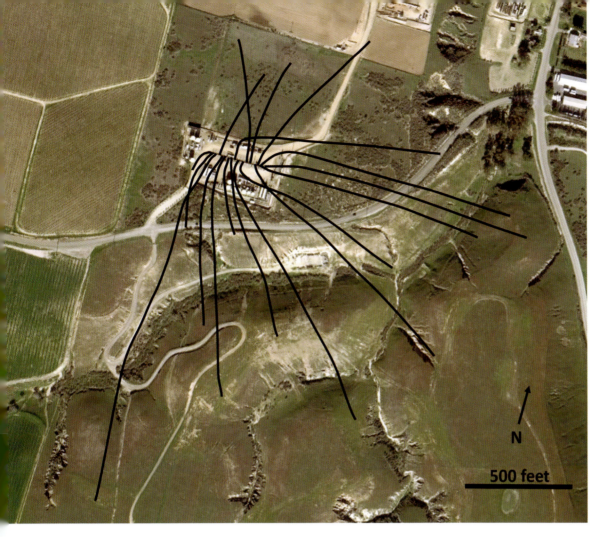

Opposite, top: Linear rod pumps (LRP) provide direct control of the sucker rod, without need for the heavy counterweight system used on conventional pumpjacks.

COURTESY OF VAQUERO ENERGY, SETH HUNTER PHOTOGRAPHER.

Left: Approximate tracklines of the directional wells, Tunnell site. The natural terrain is left undisturbed except for the small drilling site.

TRACKLINES FROM VAQUERO ENERGY; AERIAL IMAGE FROM GOOGLE EARTH.

Below: Modern pumping operation by Vaquero Energy involving a set of directional wells. Tunnell site, Sisquoc, Santa Barbara County.

COURTESY OF VAQUERO ENERGY, SETH HUNTER PHOTOGRAPHER.

N

500 feet

Aera Energy was formed from holdings of Shell and Exxon-Mobil, and is a subsidiary of affiliates of those two companies. Aera claims to have 25 percent of California's production, mostly in the San Joaquin Valley. Its proved reserves are 682 million barrels.

Hundreds of other independents have holdings ranging from just a few wells in one field to hundreds of wells in multiple fields. Some are small, perhaps family-owned concerns that have existed for decades. Others such as Warren Resources and E & B Natural Resources have grown by consolidating properties of smaller companies and majors. Many independents that operate in California also have similar mature oil properties in other states, and may be headquartered out of state. Some are diversified either locally or nationwide, or are affiliated with oil field services, refining operations, or are engaging in mining, real estate, or other industries. These companies have names like Freeport-McMoRan, Linn, Vaquero, Veneco, and Signal Hill Petroleum. Along with remaining majors such as Chevron, the independents form the backbone of today's dynamic California petroleum industry.

Although the names have changed, the history of the California petroleum industry has in some ways come full circle. Many of the issues that confronted the industry in its earliest days are still around. One of the biggest challenges is the heavy, viscous oil, up to 10 billion barrels of it still in the ground awaiting new technologies to extract it. Another largely untapped opportunity is the Lower Monterey Formation, which in many places is a reservoir rock as well as being the source rock for the oil. Although it often contains a light oil, it is also home to a peculiar low-permeability rock called diatomite from which it is difficult to remove the oil using conventional methods. Billions of barrels could be locked away in this rock. These challenges exist more or less anywhere oil is found in California.

Opposite, top: Tower to collect solar energy for production of steam. Another way to provide energy for steamflooding.
COURTESY OF CHEVRON USA.

Opposite, bottom: Steam injection pipes at Kern River.
COURTESY OF CHEVRON USA.

Above: Steam cogeneration plant at Kern River Field. This plant produces steam for steamflooding but also produces electricity for other purposes. Chevron has remained a major player in California, operating Kern River and other fields.
COURTESY OF CHEVRON USA.

Left: Two billion barrels includes what the Elwoods got with their little cable tool rig, the nation-leading output of the early 1900s, the output of the Independent Oil Producers Agency which competed with Standard in those early years, and the technology and ideas of today's Chevron.
COURTESY OF CHEVRON USA.

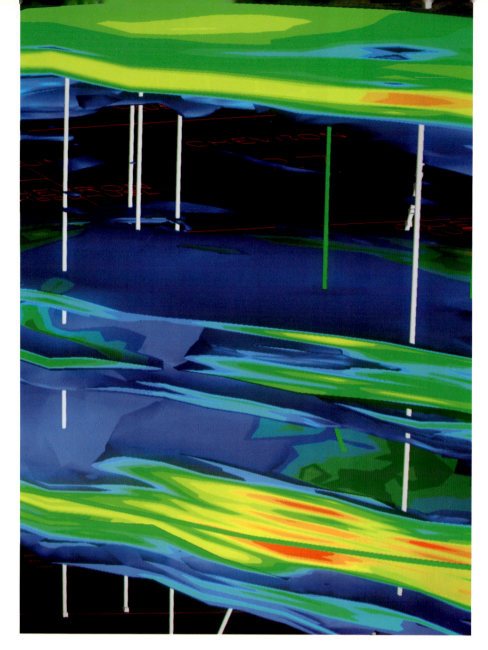

Three-dimensional model of the reservoir at Kern River.

COURTESY CHEVRON USA.

of natural gas annually. Fourteen refineries, mostly in Los Angeles and San Francisco Bay, process about 2 million barrels of oil per day. Customers receive natural gas via 100,000 miles of pipelines. Gasoline and other products are sold at 9,500 retail outlets throughout the state. In 2004 the industry had $143 billion in sales, paid $22 billion in wages and $5 billion in taxes.Its 364,000 employees account for about 2 percent of the jobs in California, a significant number for any major industry. Even with this level of production, the state still operates at an energy deficit and needs to import over 60 percent of its oil, 90 percent of its natural gas and 28 percent of its electricity.

Petroleum made California an industrial and economic giant, and made our far-flung car-oriented lifestyle possible. Although the petroleum industry has changed it will be an integral part of the California economy for decades to come. Given the growing California population and increasing demand for energy, petroleum is and will continue to be a major part of the economic life of communities across the state. Every Californian's quality of life is directly correlated with having access to ample, affordable, secure energy. The industry involves itself in local and cultural affairs and makes long-term positive contributions to California's quality of life.

America has thousands of oil companies, large and small. Each has its own strategy on how to explore for and produce petroleum. A host of technologies have been developed by entrepreneurs trying to get a competitive edge. For more than a century and a half, California has been a mainspring of this dynamic enterprise.

Today, thanks to technological advances and continued need, oil and natural gas exploration is occurring in geological locations previously unimagined. Technologies like horizontal drilling and hydraulic fracturing have shown promising results. Dire predictions of peak oil and depleted reserves have come and gone. Wallace Pratt's famous aphorism reasserts itself: Oil is, and always will be, found in the minds of thousands of explorationists in as many places. California is a well-developed petroleum province, and thanks to the legions of minds looking for oil, it has a bright future.

Information technology is also critical in enhanced recovery operations. One look at Pacific Coast Energy's control room at Orcutt Hill can confirm this. Cyclic steamflooding requires constant feedback on what is happening in the reservoir around each well. The same is true for Belridge or any other field that has hundreds, even thousands, of wells. Decisions have to be made quickly, and they must be based on reliable information. This applies to safety and environmental concerns as well as to efficient production of oil.

TOWARD THE FUTURE

Today, California produces more than 200 million barrels of oil annually from about 50,000 active wells. Although this is half of the peak production of the mid-'80s, California's domestic production is still third in the nation. The state also produces 350 BCF

MUSEUMS AND MONUMENTS: TRACES OF CALIFORNIA'S PAST

Listed here are a number of museums that have various artifacts of the early days of the petroleum industry in California, especially large objects such as pumpjacks, derricks, engines, or old-style shops and offices. Other museums may have indoor static displays that relate to the petroleum industry. In contrast, the museums listed below provide opportunities to walk among the last remaining relics of the early industry and appreciate them in a way that cannot be done with photographs.

Monuments, plaques, and other reminders can be found in many corners of the state. A few are pictured on pages 113 through 117.

ANTIQUE GAS AND STEAM
ENGINE MUSEUM

2040 North Santa Fe Avenue, Vista, California 92083
760-941-1791 • www.agsem.com

This museum has restored engines and other equipment of the type used in the oil fields, including pumpjacks, steam engines, and early hit-and-miss internal combustion engines. One steam engine was "rescued" from Pico Canyon. What sets this museum apart from the others is that the engines have been restored to the working order and appearance, down to the paint, that they would have had when they were new. These machines can be observed in operation during shows in June and October. The museum has many other types of engines, two portable cable tool rigs, early agricultural equipment and tractors, steam traction engines and other artifacts.

Above: Internal combustion engines of the type used in oilfields (arrows).
Dates of manufacture range from before 1910 to the 1950s.
PHOTOGRAPH BY THE AUTHOR, COURTESY OF THE ANTIQUE GAS AND STEAM ENGINE MUSEUM.

Left: Test engine used in research to improve gasoline.
PHOTOGRAPH BY THE AUTHOR, COURTESY OF THE ANTIQUE GAS AND STEAM ENGINE MUSEUM.

Display of cable tool components.
Left to right, Steam boiler, steam engine,
band wheel, walking beam, and base of the
derrick. The large-diameter hemp rope goes
down the hole.
COURTESY OF CALIFORNIA OIL MUSEUM,
SANTA PAULA, CALIFORNIA.

CALIFORNIA OIL MUSEUM

1001 East Main Street, Santa Paula, California 93060
805-933-0076 • www.caoilmuseum.org

Occupying the building that served as Union Oil's company headquarters before it moved to Los Angeles, this museum features a very well-preserved cable tool drilling rig. A number of gasoline pumps from early gas stations display old brands like Red Crown. The former offices of Union Oil on the second floor are meticulously restored, and one can imagine the debates between Lyman Stewart and Thomas Bard that took place in the board room. Another office was occupied by Union geologist, W. W. Orcutt, and still another was the payroll office, with enormous ledger books full of handwritten records.

HATHAWAY RANCH AND OIL MUSEUM

11901 Florence Avenue, Santa Fe Springs, California 90670
562-777-3444 • www.hathaworld.com

This five-acre museum has a ranch house, a belt-driven machine shop, and oil field equipment. Guided tours are available.

KERN COUNTY MUSEUM

3801 Chester Avenue, Bakersfield, California 93301
661-437-3330 • www.kcmuseum.org

This county museum has as one of its exhibits, "Black Gold: The Oil Experience." With indoor and outdoor exhibits of oil field equipment and other artifacts, this exhibit emphasizes the local Kern County oil industry.

MENTRYVILLE

27201 Pico Canyon Road, Newhall, California 91381
661-259-2701 • www.scvhistory.com/mentryville

This is the ghost town described in Chapter 1, with the mansion, school house, other buildings, and old equipment such as steam engines. A 1.5 mile hike up a paved road brings you to the Pico No. 4 Discovery Well and monument. A replica of a wooden derrick, old equipment, and the old pipeline can be seen along the way. The field has almost totally gone back to a natural state, but signs of it can still be seen if you look carefully.

FIRST COMMERCIAL OIL WELL
IN CALIFORNIA

ON THIS SITE STANDS CSO-4 (PICO-4), CALIFORNIA'S FIRST COMMERCIALLY PRODUCTIVE WELL. IT WAS SPUDDED IN EARLY 1876, UNDER THE DIRECTION OF DEMETRIUS G. SCOFIELD, LATER TO BECOME FIRST PRESIDENT OF STANDARD OIL COMPANY OF CALIFORNIA, AND WAS COMPLETED AT A DEPTH OF 300 FEET ON SEPTEMBER 26, 1876, FOR AN INITIAL FLOW OF 30 BARRELS OF OIL A DAY.

LATER, IN THE SAME YEAR, THE WELL WAS DEEPENED TO 600 FEET, USING WHAT WAS PERHAPS THE FIRST STEAM RIG EMPLOYED IN OIL WELL DRILLING IN CALIFORNIA. UPON THIS SECOND COMPLETION, IT PRODUCED AT A RATE OF 150 BARRELS A DAY, AND IS STILL PRODUCING AFTER SEVENTY-SEVEN YEARS.

THE SUCCESS OF THIS WELL PROMPTED FORMATION OF THE PACIFIC COAST OIL COMPANY, A PREDECESSOR OF STANDARD OIL COMPANY OF CALIFORNIA AND LED TO THE CONSTRUCTION OF THE STATE'S FIRST REFINERY NEARBY. IT WAS NOT ONLY THE DISCOVERY WELL OF THE NEWHALL FIELD BUT WAS, INDEED, A POWERFUL STIMULUS TO THE SUBSEQUENT DEVELOPMENT OF THE CALIFORNIA PETROLEUM INDUSTRY.

DEDICATED JUNE 6, 1953
STANDARD OIL COMPANY OF CALIFORNIA
PETROLEUM PRODUCTION PIONEERS, INC.

STATE REGISTERED

Top and inset: Pico No. 4 Well, with two commemorative plaques nearby. The well was plugged, and pipes left to show the well's location. An old engine and some cable for a jack line are at left.
PHOTO BY THE AUTHOR.

Left: Early oil field steam engine near the schoolhouse at Mentryville.
PHOTO BY THE AUTHOR.

OLINDA MUSEUM AND TRAIL

4025 Santa Fe Road, Brea, California 92823
714-671-4447 • www.cityofbrea.net/index.aspx?NID=438

This twelve-acre park, located on the edge of the hills of the Brea-Olinda Field, has a house that served as a field office, another building with an engine that ran jacklines going to numerous wells, the discovery well of this portion of the field, Olinda No. 1, which has been continuously producing since 1897, and a walking path that winds its way among working wells in the modern field and a few oil seeps.

Top: This building at the Olinda Museum and Trail housed the engine that ran jacklines to several wells.

Above: Jacklines to individual wells came in through the windows in the building and were attached to holes around the rim of this off-center wheel inside.
Each jackline was balanced by another one on the opposite side of the wheel. The wheel was turned by a motor via the vertical shaft, giving each jackline a back-and-forth movement.
Model of a derrick is in the background.

PHOTOGRAPHS BY THE AUTHOR.

WEST KERN OIL MUSEUM

1168 Wood Street, Taft, California 93268
661-765-6664 • www.westkern-oilmuseum.org

Located in the Midway-Sunset Field, the West Kern Oil Museum is on eight acres of land, and has the derrick and cable tool rig of the old Jameson No. 17 well, drilled in 1917. This last wooden derrick was donated to the museum in 1974, and the museum was on its way toward amassing a collection of pumpjacks, engines, boilers, old vehicles and other oil field equipment.

Above: **M**onument to oil field workers, Signal Hill. Downtown Long Beach and Santa Catalina Island in the background.

Right: **P**laques at Signal Hill. Several working wells surround a park on the steep side of the hill.

Below: **P**laque at Discovery Well Park, Huntington Beach.

PHOTOG**R**APHS BY THE AUTHOR.

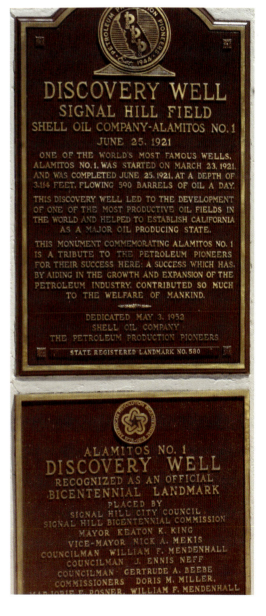

DISCOVERY WELL
SIGNAL HILL FIELD
SHELL OIL COMPANY-ALAMITOS NO. 1
JUNE 25, 1921

ONE OF THE WORLD'S MOST FAMOUS WELLS, ALAMITOS NO. 1, WAS STARTED ON MARCH 23, 1921, AND WAS COMPLETED JUNE 25, 1921, AT A DEPTH OF 3,114 FEET, FLOWING 590 BARRELS OF OIL A DAY.

THIS DISCOVERY WELL LED TO THE DEVELOPMENT OF ONE OF THE MOST PRODUCTIVE OIL FIELDS IN THE WORLD AND HELPED TO ESTABLISH CALIFORNIA AS A MAJOR OIL PRODUCING STATE.

THIS MONUMENT COMMEMORATING ALAMITOS NO. 1 IS A TRIBUTE TO THE PETROLEUM PIONEERS FOR THEIR SUCCESS HERE: A SUCCESS WHICH HAS, BY AIDING IN THE GROWTH AND EXPANSION OF THE PETROLEUM INDUSTRY, CONTRIBUTED SO MUCH TO THE WELFARE OF MANKIND.

DEDICATED MAY 3, 1952
SHELL OIL COMPANY
THE PETROLEUM PRODUCTION PIONEERS

STATE REGISTERED LANDMARK NO. 580

ALAMITOS NO. 1
DISCOVERY WELL
RECOGNIZED AS AN OFFICIAL
BICENTENNIAL LANDMARK
PLACED BY
SIGNAL HILL CITY COUNCIL
SIGNAL HILL BICENTENNIAL COMMISSION
MAYOR KEATON K. KING
VICE-MAYOR NICK A. MEKIS
COUNCILMAN WILLIAM F. MENDENHALL
COUNCILMAN J. ENNIS NEFF
COUNCILMAN GERTRUDE A. BEEBE
COMMISSIONERS DORIS M. MILLER,
MARJORIE F. POSNER, WILLIAM F. MENDENHALL

to spend the long nights with the rig crew and experience the thrill of logging a well. Early in his career, he also recognized the importance of adequate representation in politics for independent companies, and prioritized advocacy of the industry in Sacramento and Washington, D.C. Ken III has served as president of both the National Stripper Well Association and the California Independent Petroleum Association.

After so many holidays spent logging on the rigs, Ken III's son, Seth Hunter, was drawn to the oil industry at a young age. Seth graduated from Oregon State University in the year 2000, and went on to work for Halliburton in their directional drilling department. After gaining experience with a few different oil companies, he joined Vaquero in 2005 as an operator. He worked his way up to becoming operations manager in the field, and holds the title of vice president of Vaquero Energy. After years in the field and learning how the industry works first hand, Seth focuses on building a team of people that contribute to making a family run business successful through the good times and bad. He credits the company's success and expansion into four states to the stellar management teams in each region.

Seth's younger sister, Cameron Hunter, joined the company in 2013 in the regulatory and environmental compliance department. She is the first to represent the women in the Hunter family working in the oil business. She holds degrees from the University of California, Berkeley and the Middlebury Institute of International Studies in policy and business, and is focused on navigating the ever changing landscape of regulations and legislation for the oil business.

In 2007, in order to retain the unique talent of the employees by creating a growth vehicle as well as allowing those employees the opportunity of eventual equity ownership in a company, Vaquero created PetroRock LLC. The new corporation consisted of several legacy partnerships that were formed by Ken, Jr., throughout his lifetime with several undeveloped oil leases that Vaquero acquired more recently. Key employees were given shares of ownership as well as the opportunity to invest. PetroRock currently operates some of the legacy assets acquired by Ken, Sr., and Jr., and has established substantial new production since 2011 in the Cat Canyon Oilfield near Santa Maria, California. PetroRock benefits from the leadership and considerable knowledge of its president, Joe Nahama, a geologist and petroleum engineer from UCLA and the Colorado School of Mines who has been with Vaquero since 2003.

On the heels of PetroRock's success and in light of ever increasing regulations on the oil industry in California, Vaquero spun off another company based in Colorado and Wyoming called Mustang Resources LLC in 2014. The formation of Mustang, similar to PetroRock's creation, was for the purpose of retaining the talent of Vaquero's employees by allowing for eventual equity participation in a stand-alone company. Mustang Resources also represents a diversification away from oil in California, as the assets produce primarily natural gas in the Rocky Mountains region of the country. Mustang Resources is aptly led by Chuck Dobie, former president of PetroRock LLC, and a petroleum engineer with degrees from the University of Oklahoma and USC.

It is the people who Vaquero Energy and its affiliates employ that enable the business to continue through an ever-changing business climate. From its beginnings in 1942 as a handful of assets purchased, operated and traded by Ken, Sr., to a foothold in Bakersfield, Ventura, and Santa Maria in California; West Texas; Wyoming; and Colorado in the rest of the country, Vaquero continues to expand and seek new opportunities in the oil and gas industry. The Hunter family is proud that the company has persevered as a family-owned entity, and they are grateful for the opportunity to operate a business they love alongside talented colleagues and associates.

Above: Early construction days of the PetroRock Tunnell Facility in Santa Maria, California, in 2012 after a rainstorm.

Below: Left to right, Erik Vasquez, Seth Hunter, Ken Hunter III and Joe Nahama of PetroRock in 2014.

BENTLEY SIMONSON, INC.

Right: C. T. Simonson (third from left) Clark #6, Santa Fe Springs, California, c. 1923.

Below: Simonson's uncle (second from left), Los Angeles, California, c. 1925.

The rich history of the California petroleum industry is replete with stories of daring entrepreneurs, fearless wildcatters and swashbuckling, larger-than-life individuals who achieved great success through their vision, intelligence, and endless hours of unstinting labor.

This colorful history would not be complete without including the achievements of Clif Simonson, who rose from humble blue collar beginnings to establish and operate dozens of successful oil field ventures.

Simonson is a 'born-in-the-business' third-generation oilman, the descendant of men whose careers dated from the 1920s and includes work on such legendary fields as Getty Oil's Ventura Avenue field to today's greatest finds such as the Haynesville and Bakken. His success is rooted in this history.

The Simonson family moved to California from Nebraska, by way of Colorado, in the early 1920s, after grandfather O. T. Simonson found little success in multiple ventures, including a livery stable that mysteriously burned down after a dispute with his partner. "O. T. was a hell raiser, but also an entrepreneur. As my dad put it, his mistake wasn't in the hell raising or the work, but in the ratio—less hell raising and a hell of a lot of work rather than the reverse is better." Simonson explains.

In the 1920s the oil business in Los Angeles was booming, and Simonson's father—Carlton "C. T." Simonson—had no trouble finding work on a drilling rig. Simonson's first picture

of his dad in the oilfields was taken on the floor of a drilling rig in 1923. In the early 1940s, due to an eye injury, C. T. moved from drilling to production where he worked as a field operator, roustabout foreman and field superintendent. But his growing family required him to do more and, like his father, he exhibited a strong entrepreneurial streak. He learned from O. T.'s experience and did it with less hell raising and a hell of a work ethic. Both of these traits he passed onto Simonson, along with the knowhow of years in oil and gas production.

"He had only about a sixth grade education, but he did a lot of stuff other than his day job," says Simonson. "In his off time he did things like build houses and real estate development. We also had a small farm where we raised our own cattle, chickens and pigs.

"And just like my dad's father, my other grandfather on my mother's side was also a ne'er do well," Simonson relates. "However, with my dad's assistance, this grandfather pulled his life together after Simonson's father got him a job in a field Getty had drilled in Oxnard. It was there that young Simonson's love of the oil business was born.

"It was a relatively small field and my maternal grandfather lived in a trailer on the lease in the early 1960s and looked after the field," Simonson recalls. "He was a good grandpa and when I was nine or ten years old, I was out there with him, greasing and oiling things, and gauging the tanks."

C. T. retired in 1969, about the time his father-in-law developed lung cancer and was forced to quit work. "My grandfather passed,

and my dad's retirement job was looking after oil wells, so all through elementary and high school I would ride around with him and help take care of that particular field and others."

At the same time C. T. also partnered with Simonson's half-brother, Bob, and formed D&S Welding (later D&S Industrial Services), which provided welding and manufacturing for the oil fields. C. T. also expanded into contract pumping. "My brother, Carl (now a petroleum engineer for Chevron) and I worked for my dad as roustabouts and field mechanics. We were cheap labor. Bob was often my boss. He was a true oilfield SOB but I learned a lot from him. I also welded and ran heavy equipment," he explains.

In 1979, Simonson started college at Chico State in Northern California, where he worked toward an unconventional double degree in English Literature and Business. Tragedy struck the family during Simonson's senior year at Chico when his younger sister died in a motorcycle accident. Then, his father was killed in an auto accident six months later. "It was a dark year, a tough Christmas" he recalls. "Afterwards, I went back to Chico, packed up my stuff, and came home to take up the family business." Although he regretted dropping out of college in his senior year, he later went to night school and received his business degree.

Simonson, twenty-one years old at the time, grew the business by focusing on contract pumping. During the next decade the company expanded into Bakersfield and added additional services such as environmental consulting, engineering, and construction.

D&S also grew by devising a unique method of acquiring the management of oil fields. The industry, always subject to boom and bust periods, was in a down cycle around 1986 and lots of oilfields were on the market. "I would cold call both the sellers and the buyers of these properties and ask who was going to take care of their oil wells. It was to both their benefits to have a quality operator for the property," he explains. "They usually hadn't thought that far ahead, but from past experience I usually knew the fields and I could give them a bid on the spot. It wasn't long before we put most of our competitors out of business."

Managing the oil wells included providing the maintenance and handling of complicated regulatory matters. The company developed a unique bookkeeping system for tracking the projects and Simonson developed a software program to track production and provide regular production reports.

Simonson also discovered he was working constantly. "We had this time-keeping system where we kept track of every ten minutes of our time, just like a lawyer," he recalls. "I was looking through a computer printout of my time one day and realized I had worked seven straight years with no days off. That included Christmas, New Year's and 4th of July. It was 365 days times seven years. And there were times when I worked two days straight—forty-eight hours. It began to dawn on me that I was

Above: C. T. Simonson, Dressing Tools, c. 1935.

Below: C. T. Simonson, Getty on Ventura Avenue Oil Field, 1969.

working all these hours but not really getting ahead. I knew I had to do something different."

Because of this effort, by 1990 the company was doing well but Simonson was looking for new business challenges. In 1987 he had partnered with Ted Bentley, forming Bentley Simonson, Inc. (BSI), and they had purchased a few oil properties. "It was a small venture," Simonson explains. "I realized I was making a living, but not building any real wealth. Buying more and bigger oil fields under the BSI banner, seemed like the way to accomplish this, but I needed to learn a lot more about the business of the oil business.

"Around this time, I cold called a company, following up on a lead from the property divestiture department at Unocal. The company, Saba Petroleum, was buying fields from Unocal. Before I even called them I

realized that they needed to buy my company as they had no way to operate their new assets. I had twenty field operators and an oilfield construction company. I needed some cash to buy out a partner. We negotiated a deal that included some cash, but mainly stock. Very cheap, pink sheet traded stock. It was a huge risk. I did the deal in thirty minutes on the phone. It was a life changing call."

Simonson joined the corporate world as vice president of Operations for Saba Petroleum. "I was on the front lines of setting up operations of the oil fields Saba was acquiring," he says. This included establishing administrative systems, and hiring operations, technical and management personnel. It was the same work he had done for his former clients, but on a much larger scale. He relished the challenge. There was one problem, "The company was owned by a group of Pakistanis so there was a bit of culture shock and adjustment to their way of doing business. However, their big deal was the promoting of the company stock. Which was good and bad. Good for the stock, but as we grew I felt expedient decisions were made that would negatively affect the long term health of the company. The first two years were great, but then it became a political hellhole. And fundamentally, I realized I had always been my own boss, and that I couldn't adjust to working for someone else."

Simonson left Saba in 1994 and began once again to flex his entrepreneurial muscles. He built up and sold an environmental services

company. He became a student of the stock market because of his ownership of his former employer's pink sheeted stock. (His wife was a stockbroker, which helped tremendously.) It was during this time that Saba stock had become a high flyer, and began trading on the American Exchange. However, Simonson realized the damage being done to the long term prospects of the company by management's short term focus. He sold his shares at the top and became a millionaire, at least for the moment. (Two years later his instincts were confirmed and Saba ceased to exist.) He was still a partner in BSI, which, since 1987, had owned the Maulhardt lease, the property his grandfather had looked after years before, as well as several other small oil properties.

He then invested heavily in the Maulhardt field. "We decided to drill some wells. The first one was a boomer. So, we decided to drill two more. However, there were mechanical problems on the third well. Basically, we lost the well," Simonson explains. "At the same time the price of oil had dropped and we were done—broke. Now I was back to square one. It was devastating."

At this point, Simonson was in his late thirties and, in effect, was starting over. A lawsuit over the botched drilling took seven years of litigation and although Simonson eventually won a big judgment, the lawyers got most of the money. "The suit didn't help financially, except that I received a priceless legal education," he comments. Now armed with the experience as an

entrepreneur and in the corporate world, and the expensive education from the school of hard knocks, Simonson decided it was time to revisit the vision of building out the Bentley Simonson brand. Rather than drill wells, he wanted to buy existing fields and rehabilitate them. He needed help. Specifically, he needed a petroleum engineer to assist in evaluating properties.

By happenstance, Simonson was visiting a local Kinkos copy shop one day and met a petroleum engineer from Norway, Petter

Top: Oxnard Oil Field, c. 1990.

Above: Clif Simonson (left) aftermath of a field repair on a leaky value, c. 1990.

Left: Clif and Andrea Simonson with daughter, Olivia, drilling first well, Oxnard Oil Field, c. 1997.

Above: Original founders of Bentley Simonson (left to right) Jim Bentley, Ted Bentley, Clif Simonson, and Petter Romming, East Los Angeles Field, c. 1998.

Below: Haynesville Gas Well, Red River Bull Bayou Field, Louisiana, 2010.

Romming, who was working on his resume. The two hit it off and were soon talking about collaborating on various projects.

Soon Simonson and Romming were driving all over the Los Angeles area, looking for worn-out properties that might be salvageable. "We finally found this really messed up oil property called the East Los Angeles Field," Simonson says. "It was contaminated and all the power had been shut off and there were no permits and everything had been red-tagged by various government agencies. There were title issues as well. But I like a challenge and it was a great place to learn because it had every type of problem you could imagine."

Working with his main partners, Bentley, and Romming, the team undertook a two

year plan to rehabilitate the property. Bentley's brother, Jim, came on board as construction foreman and a number of wells were returned to production. "We went from zero production to 300 barrels of oil a day, which is not too bad," Simonson notes.

Pleased with the success of the rehab, the partners decided to do a second rehab project in 2000. Acquisition of the first project had cost only $5,000, the second carried a price tag of $5 million, which was borrowed from Union Bank. The Las Cienegas Field consisted of a series of urban drill sites in Los Angeles. The property was in bad shape before it was rehabbed. The team was able to increase production from 300 barrels a day to over a thousand barrels a day in just under two years.

Other rehab projects followed and the company soon had more than a hundred capital projects going and more than 100 employees. They had low debt and good reserves, which made the company very valuable. But finances were still very tight and Simonson's wife had been diagnosed with cancer.

At this point, the partners felt they needed to decide whether to continue expanding the business or put it up for sale. A consulting firm was retained to analyze the situation and make recommendations.

"It soon became clear that our company was worth a lot of money, but not everyone agreed," Simonson says. "We were on a conference call with our banker one day when I mentioned

that the company should sell for more than $100 million. The banker responded with a barnyard epithet and stated that he would kiss a dog's behind if the company sold for more than $100 million."

The company's oil and gas assets were sold in April of 2005 for $119 million. At the closing celebration, one of the lawyers brought in a huge stuffed dog so the banker could make good on his vow. He declined, however, and is now a partner and executive with one of Simonson's companies.

The success in purchasing "broken-down" oil companies and rehabilitating them so they became profitable again served as a springboard for a number of new ventures for Simonson and his partners, including real estate, service companies, and, of course, oil and gas exploration.

It was during this time that Simonson dialed back his work schedule and devoted time to looking after his wife and their two children, taking her to and from chemotherapy and doctor's visits, and traveling with her when she felt well enough.

Andrea Grace Soter-Simonson died in 2008. In her memory, Simonson established the AGSS Memorial Foundation to aide families struggling with cancer. To date, the Foundation has helped several hundred families in Ventura County.

From the remains of the old BSI, a holding company named True Oil was founded to establish and operate new ventures. Among those new firms were BSI Energy Partners and Simonson's own private equity company, The Ceson Group.

The company also continued to look for "broken-down" oil companies that could be rehabilitated. These included a fix-it-up property in Louisiana called The Red River Bayou. As it turned out, the property was sitting on the bull's eye of the largest gas field in the nation.

Simonson has also been involved with California Well Services and North Dakota Well Services. These companies provide maintenance service to the oil industry. "With our affiliate—North Plains Energy—we were in the Bakken early and thus fortunate to participate in one of the biggest oil booms in decades," says Simonson.

Currently, Simonson and his partners are involved in about twenty diverse companies, with a core group that remains connected to the oil industry. Still a relatively young man, Simonson is looking forward to even more creative and successful ventures in the future.

"I think I've shown that hard work really does pay off," Simonson says. "It continues to be a team effort. I couldn't have done it without my partners. I've been the torch bearer and head cheerleader but we've all contributed to the successes along the way. I love what we do and feel I am just getting warmed up."

North Dakota Bakken, first production 2011.

CALIFORNIA RESOURCES CORPORATION

California Resources Corporation (CRC) is the largest oil and natural gas producer in California on a gross-operated basis. The company has established what is believed to be the largest privately-held mineral acreage position in the state, consisting of approximately 2.3 million net acres spanning California's four major oil and gas basins. CRC was spun off from Occidental Petroleum Corporation in 2014.

Headquartered in Los Angeles, CRC and its employees and contractors are committed to providing Californians with long-term ample, affordable and reliable energy exclusively from California resources.

The local oil and natural gas industry allows Californians to consume energy that is produced in California, by Californians, for Californians. Energy self-sufficiency that reduces our reliance on foreign oil means greater economic and national security, less environmental impact from shipping oil across long distances, and jobs and revenues that are reinvested in our local economy and people.

CRC operates in more than 130 fields in California, including its flagship Elk Hills and Wilmington Fields, which were discovered about a century ago. Most of CRC's assets were acquired or built over the past eighteen years through targeted investments in California.

CRC was formerly part of Occidental Petroleum Corporation, which was founded in California in 1920. On February 14, 2014, Occidental announced that it would spin off all its California operations into a separate, publicly traded company. On November 30, 2014, CRC was officially an independent company.

Todd Stevens is president and CEO of CRC. Previously, Stevens served as vice president, Corporate Development at Occidental Petroleum Corporation, where he led growth-focused initiatives, including mergers and acquisitions, land management and worldwide exploration. Prior to that, he served as Occidental's vice president, Acquisitions and Corporate Finance, as well as vice president, California Operations. During his twenty years with Occidental, he managed or was involved in transactions totaling more than $18 billion.

CRC has operations at the Elk Hills Field in Kern County, the Wilmington Field in Long Beach and fields across the Los Angeles, San Joaquin, Ventura and Sacramento basins.

More than ever, Long Beach is a major player in Southern California's coastal economy, bringing together industries from tech to trade to energy that benefit city residents and all Californians. The roots of oil and gas-driven prosperity for the people of Long Beach run deep. Much has changed in Long Beach over the years. Nonetheless, the safe production of oil and natural gas that helped lay the city's foundation continues to be a core element of the local economy.

The Wilmington Oil Field was first discovered, partly under Long Beach and its Harbor, in 1932. In 1962, local residents approved an innovative plan for tapping into the rich Wilmington Field in Long Beach Harbor, all while upholding the area's high environmental standards. Conditions of the field development included:

Left: Sunset Field, Kern County, California.

Right: Lost Hills, General Petroleum Company tanks and derricks.

- Field had to be operated as a single unit with sustained water injection,
- City of Long Beach was to be the unit operator with control over operations, and
- Development was to be confined to the pier areas and oil production islands with architecture and landscaping consistent with the coastal development.

The Wilmington Field runs from southeast to northwest, beginning at a point offshore of Seal Beach and ending underneath Redondo Beach. The oil field consists of large anticline, or underground dome, made of layers of permeable rock containing tar and relatively-heavy crude oil. Because of this structure, advanced recovery techniques are required to remove the bulk of the available oil.

In 1965, development of a project known as THUMS began under direction of five oil companies who bid for the right to develop and produce oil from under the City of Long Beach and Harbor. Texaco, Humble (now Exxon), Union, Mobil and Shell established THUMS Long Beach Company, whose name

(THUMS) was derived from using the first initial of each of the founding five oil companies submitting the successful bid.

The first THUMS well was started on July 16, 1965, on Pier J, near the present site of the *Queen Mary*. Within three weeks, the well was producing more than 700 barrels of oil per day. By the end of the year, nineteen wells were producing 11,500 barrels per day. But the long-term plan called for most oil production to come from further out in the harbor, so construction began immediately on the four oil islands. Originally, the THUMS islands were designated simply as Islands 'A', 'B', 'C', and 'D'. They were later renamed the Grissom, White, Chaffee, and Freeman Islands in honor of the astronauts who died in the U.S. Space Program.

As the oil business evolved, each of the five original THUMS stockholders sold their shares to a subsidiary of Atlantic Richfield Company (ARCO). ARCO invested in the future of THUMS and the Long Beach Unit by implementing plans for $100 million in exploration and development of the unit.

Four views of THUMS islands from initial construction in 1965 to the present.

Bottom, right: CRC's operations at THUMS Island White in Long Beach, California 2014.

Above: Pumping unit and drilling rig at CRC's Elk Hills Field near Bakersfield, California.

Below: Drilling rig, Sacramento Basin.

In 2000, ARCO was acquired by BP and THUMS was purchased by Occidental Petroleum Corporation, which continued with CRC to invest heavily in Long Beach.

Since it first came into existence, THUMS has pleased local residents by the skill and diligence demonstrated in protecting the environment and enhancing the beauty of the harbor and beach front, keeping with the company's commitment to the City of Long Beach.

Residents and tourists seldom recognize the THUMS islands in Long Beach Harbor as centers of round-the-clock industrial operations. Instead, they are viewed as attractive residential or recreational areas. The oil production operations on the islands are thoroughly camouflaged, sound-proofed or placed underground.

The Grissom and White Islands are made even more attractive by specially designed drilling towers. The conventional oil derricks and pipe racks are covered with attractive, sound-dampening insulated panels that many feel give the towers the look of high-rise condominiums or apartment buildings. No more than four towers have operated at any one time on any island. Each is moved from location to location as new wells are drilled.

The Long Beach Unit comprises about forty percent of the total Wilmington Field and was created by combining three adjacent mineral tracts. The Long Beach Unit extends under about 6,500 acres of land and water. The four islands are evenly spaced above the Long Beach Unit to provide efficient drilling distances to the major oil reservoirs.

Since 1965, THUMS has worked to maximize production from the oil fields located under the Harbor and the City of Long Beach, while establishing an enviable record of safety and environmental protection.

THUMS has produced more than 1 billion barrels of oil equivalent in total from the Long Beach Unit. This partnership of government, public and private entities has resulted in more than $4.3 billion in revenue for the City, State and Port of Long Beach.

CRC operates fifty-three fields in the Sacramento Basin primarily consisting of dry gas production. CRC currently holds approximately 0.5 million net acres in the basin, approximately thirty-five percent of which are held in fee. CRC is the number one gas producer in the basin and believes its significant acreage position in the Sacramento Basin gives the option for future development and rapid production growth in an attractive natural gas price environment.

The Sacramento Basin is a deep, elongated northwest-trending basin located in northern California covering about 12,000 square miles and forming the northern part of California's Central Valley. It contains a thick sequence of sedimentary deposits that range in age from lower Cretaceous to Neogene sediments in an area that is approximately 200 miles long and forty-five miles wide. Producing reservoirs range from upper Cretaceous-age to Pliocene-age. The main reservoirs are the Cretaceous Starkey, Winters, Forbes, Kione, and the Eocene Domengine sands. Exploration in the basin started in 1918 and was focused on seeps and topographic highs. In the 1970s the use of multifold 2D seismic surveys led to large discoveries in the basin. The acquisition of 3D seismic surveys in the mid-1990s helped define trapping mechanisms and reservoir geometries. The Sacramento Basin has been extensively explored for petroleum resources, and more than 10 trillion cubic feet of natural gas have been produced.

If Kern County were a state, it would rank fourth in oil production—and it is the leading county in oil production in the lower forty-eight states. Kern County's abundance of energy production powers the state, bolsters the local community and keeps energy affordable for California families.

The giant Elk Hills Field is a key factor in CRC's position as the largest natural gas producer and the largest oil and gas producer in California on a gross-operated basis. Located in the southern portion of California's San Joaquin Valley, Elk Hills is one of the largest oil and natural gas fields in the United States.

The Elk Hills Field, which covers seventy-five square miles, was discovered in 1911 and has produced more than 1.6 billion barrels of oil equivalent (BOE), thus making it one of the most productive fields in the nation. In 2014, CRC produced 64,000 BOE per day on average from the Elk Hills properties. As California's number one gas producer, production from Elk Hills' more than 3,000 active wells contributes over forty percent of California's natural gas production and approximately five percent of the state's oil production.

CRC's operation of the field began in 1998 after winning the bid for the historic Elk Hills Naval Petroleum Reserve near Bakersfield and paying $3.65 billion. This was the largest sale of U.S. Government property to that time.

Elk Hills exemplifies CRC's focus on acquiring large, long-lived properties and applying enhanced oil recovery techniques to increase production in mature fields. CRC has customized a range of recovery techniques to the field's complex reservoirs, transforming this century-old field into the largest natural gas producer in California.

Elk Hills, which includes an 8,000 acre conservation area, has been recognized for outstanding environmental stewardship by the State of California and such safety and environmental organizations as the Wildlife Habitat Council, California Climate Action Registry and the National Safety Council.

Above: Pumping units at CRC's operations in Ventura, California.

Below: Pumping unit at CRC's Elk Hills Field near Bakersfield, California.

EXPLORATION & PRODUCTION

Above: Enclosed drilling rig at CRC's operations in Huntington Beach, California.

Below: Drilling rig at CRC's Elk Hills Field near Bakersfield, California.

Elk Hills programs and initiatives include the preservation of wildlife habitat and cultural resources, emissions reduction, waste minimization and recycling.

CRC actively operates and is continuing to develop twenty-nine fields in the Ventura Basin, which consists of primary conventional, improved oil recovery (IOR), enhanced oil recovery (EOR), and unconventional project types.

CRC currently holds approximately 0.3 million net mineral acres in the Ventura Basin, about seventy-two percent of which are held in fee. As of December 31, 2014, approximately eight percent of CRC's estimated proved reserves and six percent of the company's average daily production were located in the Ventura Basin.

The Ventura Basin is the onshore part of the main structural feature and its offshore extension is the modern Santa Barbara Basin. All of the sedimentary section is productive at various locations and most reservoirs are sandstones with favorable porosity and permeability. The basin contains multiple stacked formations throughout its depths and CRC believes the Ventura Basin provides an appealing inventory of existing field redevelopment opportunities, as well as new play exploration potential.

CRC and its employees are very involved in their local communities. CRC believes that contributing to the communities where its employees live and work is both a business and an individual responsibility as Californians.

CRC serves as an active and supportive community partner through impactful donations and volunteerism. CRC's charitable contributions support CRC's five charitable giving pillars:

• Veterans—CRC supports military and veterans organizations in recognition of their service to our country and as valued members of our workforce.

• Education and Job Training—CRC supports programs that inspire students to learn about science, technology, engineering and mathematics (STEM) and the oil and natural gas industry, ranging from hands-on learning and after-school opportunities for elementary school students to job training partnerships, internships and scholarships for young adults.

- Public Health and Safety—CRC supports and develops programs that promote community health, safety and well-being, focusing on children's health and wellness for communities in need and our public safety professionals.
- Civic Events—CRC is committed to the communities where its employees live and work. Reflecting the diversity of its workforce, CRC supports community programs and events that strengthen California's vibrant neighborhoods and address the local priorities that make each distinct.
- Environmental Stewardship and Water Conservation—CRC has a dedicated 8,000 acre habitat conservation area at its Elk Hills Field, which is home to multiple animal and plant species; CRC supports environmental efforts led by local organizations to promote habitat conservation and biodiversity such as in the Bolsa Chica wetlands in Huntington Beach; and the company partners with water districts and nonprofit organizations to help alleviate the impact of California's drought on farmers and rural communities.

Sensitive to its role in conservation, CRC's steam flood operations in the San Joaquin Valley during 2014 supplied more than two billion gallons—more than 6,200 acre feet—of treated, reclaimed water for irrigation. CRC's operations in Long Beach use recycled water for approximately ninety-nine percent of their total water use.

CRC's business planning for the future is focused on delivering long-term shareholder value by applying modern technology to develop the company's resource base and increasing production and recovery factors. With significant conventional opportunities to pursue, CRC uses life-of-field planning to increase recovery factors by transitioning from primary production to steam floods, water floods and other enhanced recovery mechanisms. CRC is committed to living within its cash flow, and is prioritizing oil projects that provide low production declines and high returns, such as steam floods.

President and CEO Todd Stevens praised CRC's workforce in December 2015 stating: "We are extremely proud of our team's accomplishments during our first year as a stand-alone company. In a challenging commodity environment, we increased our production despite spending below the level of investment we expected would be required to maintain production. This production increase is a testament to our employees' effective management of our low-decline asset base. The attributes of our world class assets provide us with increased flexibility to withstand a prolonged downturn in commodity prices."

The company's long-term business strategy consists of four key elements:
- Focus on high-margin crude oil projects to generate cash flows to fund the capital budget internally;
- Increase the share of conventional projects in the production mix to achieve lower declines and lower capital requirements;
- Develop high-growth, unconventional drilling opportunities and mature its exploration portfolio; and
- Continue to prioritize safety, environmental protection and community relations.

California homes, farms, businesses and communities need ample, affordable and reliable energy, and California Resources Corporation's dedicated workforce will continue to provide "Energy for California by Californians" by developing dependable local crude oil and natural gas reserves that help to reduce the state's chronic dependence on imported energy.

Platform Emmy, CRC's offshore platform in Huntington Beach, California

Reward Lease gusher near Taft.

Clarence mined for gold on the Fortymile River, a tributary of the Yukon River in Canada, and worked the mines for eighteen months before returning home to marry his childhood sweetheart, Ethel Bush. Eager to return to Alaska, the newlyweds—along with Clarence's younger brother, Fred— caught a boat to Alaska and then trekked north by foot and dog team to the mining town of Fortymile.

By then, the area had been combed over by many prospectors for a decade and the Berrys' had little luck in finding gold. His funds depleted, Clarence was forced to work as a bartender in a saloon operated by Bill McPhee. Clarence was practically broke when the news came of the Yukon gold strike that would become one of the richest gold discoveries in history.

Eager to take advantage of the once in a lifetime opportunity, Clarence borrowed the money from McPhee and quickly outfitted an attempt to stake a claim in the new gold fields. Wasting no time, Ethel was dispatched to hail a boat headed upriver while Clarence

and his brother gathered the food, clothing, tents and tools needed for the venture. Because of their quick response, the Berrys' were among the first to reach the Yukon and stake a claim on Eldorado Creek. That claim soon made them wealthy. Clarence never forgot the help he received from McPhee and remembered him in his will.

By 1902, Clarence had mined $1.5 million from his Eldorado claims and would continue to mine them for another ten years. He went on to make two more fortunes in Alaskan gold, first near Fairbanks and then around Circle City, where he mined for the rest of his life.

"Gold would bankroll many a business venture for the Berry family back in California. But there was another fortune waiting underground closer to home that would dwarf all their holdings, even the gold mines," writes Lumbye.

By the time Clarence left Alaska in the late 1800s, Kern River, along with the nearby Midway-Sunset Field, had made California the leading oil producing state in the nation. With the wealth he had accumulated in Alaska, Clarence was able to buy up promising tracts of land in Kern County in 1899 and soon began drilling for oil. In 1909 he completed his first successful well in the Midway-Sunset Field and created the Ethel D. Company, named after his wife. That well is still producing oil to this day.

"Following the same *modus operandi* he'd employed in the gold fields, Clarence cast a wide net in the oil business, buying property and forming new companies and partnerships. He'd keep it up for the rest of his life," writes Lumbye. Clarence had been very successful as a wildcatter in the late 1800s, but even in the 1920s, oil remained a volatile business.

"It wasn't just that he loved new ventures, although he clearly relished the thrill of the launch. To Clarence's way of thinking, there wasn't much point in being rich if you couldn't bring your friends and family along. Each new entity became a way to share the wealth. If his brothers didn't have enough money to buy in, he'd advance it, with no real intention of foreclosing if they never paid up."

Clarence was based in San Francisco, but operated far and wide. Newspaper stories of the day have him traveling to Arizona to buy cattle, investing in a gas-powered sled in Alaska, and helping start a bank in Selma. He bought ranches and wheat fields in the San Joaquin Valley and took time to hunt deer in the Sierra foothills and go duck hunting in the wetlands.

Realizing the proliferation of companies was becoming unwieldy, he formed Berry Holding Co. in 1916 to pull all the companies together. By 1925, Clarence was either a founder or major stockholder in Berry Holding Co. and at least nineteen other companies, including a dredging company, he formed in Alaska.

In a memoir he never got around to finishing, Clarence wrote, "I could have made a lot more money had I tried, but what good would it have done me? A certain amount is necessary or rather handy, but after looking around at some of my friends and being familiar with their habits and seeing them make slaves of themselves, all for what? They don't mean to do it, but they become money mad, not necessarily for the use they get out of it, but to outdo their neighbors. They certainly are to be pitied. Most of them after making a lot of money feel above the ordinary person. I could never see why money made any one better. In some cases, where they try to help the needy, yes. In many cases they became selfish, disagreeable and sad for the want of friends, which money will not buy."

Clarence died unexpectedly of a ruptured appendix in 1930 at the age of sixty-three. His last words, whispered to his nephew, Othmar Berry, were, "Keep it going."

Determined to follow his deathbed wishes, members of the Berry family continued to operate the companies. In 1967, as the second generation of the family began to age, a member of the third generation, Peter Bennett, a journalist by training, was asked to step in and guide the company.

The company was revitalized under Bennett's direction and he continued to head the family firm until 1983, when professional managers were brought in, led by Ray Bradley and Sam Callison, followed by Harvey Bryant.

The company enjoyed an unexpected windfall during the Arab oil embargo of 1973 when gasoline shortages resulted in long lines at gas pumps across the nation. Prices more than quadrupled to dizzying heights of five to ten dollars a barrel for oil produced after the embargo.

Above: C. J. Berry (center) and friends in front of a lake of oil, on the Ethel D. Lease at Midway-Sunset Field, 1909.

Below: Ethel D. Lease gusher and lake of oil, 1909.

Above: C. J. Berry panning gold with wife, Ethel D., standing on the far right. Henry Berry and his wife Tott are on the left, Klondike Yukon, 1898.

Below: Clarence Berry and wife Ethel in front of Ethel D. Lease, 1910.

The Berry companies now had a new challenge—what to do with all that money. "Now we had the responsibility not only of producing, but managing the new wealth," explained Bennett. "Now, suddenly, we had to manage our dividend policy more effectively, more generously, and stabilize it; and there would be a considerable amount of money left for something else. We had to figure all this out on the run."

The windfall led to two major acquisitions that had very little to do with the oil business, the purchase of a struggling Long Beach shipyard and a high-end sheepskin tannery. Neither venture proved very successful.

In 1985, in order to streamline operations, Berry Petroleum Company was incorporated in Delaware. It became the surviving company after merging with Berry Holding and a number of other family enterprises. Eagle Creek Mining and Drilling Company was created to hold non-oil and gas producing assets, including a well servicing and drilling company.

The next step in the transformation of Berry was the purchase in 1986 of an eighty percent stake in the Norris Oil Company. The balance of the company was acquired in 1987.

Through the Norris purchase, Berry picked up oil and gas properties in the Rincon Field in Ventura County. Also through the Norris Purchase, Berry became a publicly traded company, available over the counter through NASDAQ. Nevertheless, the company remained very much a family-controlled business, with at least three-quarters of the company's stock in the hands of Clarence's descendants.

Berry continued as a relatively small but profitable company through the 1980s, although it was at the mercy of the major oil companies in transporting its Kern County crude to California refineries. Conditions improved somewhat with the opening of the All-American pipeline.

The worldwide oil glut of the early 1990s forced the major producers and many independents to shut down their wells. Berry, with its lower cost structure, was still able to turn a profit, although 400 of its 1,300 wells were shut down.

Berry essentially broke even in 1993 but lost $1.1 million in 1994. The company would have actually turned a slight profit in 1994 had it not incurred a $1.3 million charge related to an oil spill in December 1993, an accident that resulted in bad publicity for the company.

Berry returned to profitability in 1995 and management initiated a five year growth

strategy that resulted in the acquisition of a number of valuable oil properties in the South-Midway-Sunset Fields as well as three cogeneration facilities. As a result, Berry's reserves increased significantly and the company was well positioned to increase production to take advantage of oil prices that rebounded in 1999.

The energy crisis of the early 2000s had an adverse effect on Berry when two of its major customers for electricity—Pacific Gas and Electric Company and Southern California Edison Company—were pushed to near bankruptcy and were unable to pay for the power Berry had provided. Berry was forced to shut down four of its five turbines and curtailed its capital development program, but despite these setbacks, the company enjoyed another profitable year.

The power situation in California finally stabilized and because of its cogeneration plants, 100 percent ownership of most of its oil-bearing properties, and financial stability, Berry was well positioned in the early years of the twenty-first century. In 2003, Berry started exploring the Uinta Basin in northwestern Utah.

The family controlled company became an attractive acquisition target and in December 2013, Berry Petroleum Company was purchased by U.S. oil and gas producer Linn Energy LLC. The company sold for $4.9 billion.

Berry family members could only imagine what Clarence, who started searching for his fortune in Alaska with only $3.60 in his pockets, would think of the empire he built on guts, sweat, swagger and almost unimaginable luck.

Special thanks to author Betsy Lumbye, whose biography of the Berry family—Beyond Luck— was indispensable in the preparation of this profile.

Above: Tott and Ethel Bush Berry are sisters married to two Berry Brothers, in Eldorado Creek Sluice, 1898. C. J. is shoveling nuggets.

Below: Surprise Oil Company building on Hovey Hills Road, Taft, 1926. The building is still there; it is now the Linn Energy LLC.

HOWARD SUPPLY COMPANY

Above: Old photograph of a Howard Supply Company truck, Los Angeles, California.

Below: The former Howard Supply Company store in Taft, California.

Opposite, top: Left to right, Vice President Wayne Moody, Western District Manager Tom Work, AESC National President Ken Gates and AESC Executive Director Kenny Jordan at 2005 Wire Rope Shop dedication.

Opposite, bottom: Part of the crew at the Bakersfield Store. Luis Lopez, second from right is now the Western Area manager.

An ability to change with the times and find its proper niche has enabled Howard Supply Company (HSC) to grow and prosper for seventy-eight years. Over time, a heritage of excellence and service has made HSC the premier value added supplier to the well-servicing segment of the oil and gas industry.

The company was founded in Los Angeles in 1937 by Harold Howard, who grew the firm from a single supply store into a large organization with thirteen branches throughout California. The Howard family continued to own and operate the company until 1977, when it was sold to Republic Steel Company.

In those days the 'Big Three' steel companies each owned a supply company that became the distribution arm for its products. Thus, Howard Supply became the distributor of Republic Steel products, along with the many other items needed in the oil fields.

A merger of Republic Steel and LTV Steel in 1983 created a situation where the merged company owned two supply companies. Republic Steel merged with LTV Steel in 1983, creating a single company with overlapping territories and two supply companies. As a result, Republic—including Howard Supply—was put on the market and purchased in 1985 by Frank M. Late. Late, a Texas native, owned a number of companies in the oil fields, including Cactus Drilling. When Late acquired Cactus it operated two drilling rigs; when he sold it in 1981 there were more than a hundred in operation.

When it became an independent business in 1986, Howard Supply was a relatively small supply company with two stores in Bakersfield and Taft with about twenty employees. "At the time, HSC was very well known for its expansive product lines; we carried everything you could imagine for the oil field and our catalog looked like a Sears-Roebuck catalog," according to Wayne Moody, former president and CEO. "In the early days, if you wanted to buy pipe, if you wanted to get a derrick, if you wanted a pumping unit—virtually every tool or pipe, valve fitting instrument, power transmission, anything you could think of—Howard Supply had it."

For a time, HSC was even heavily involved in the construction of gasoline service stations. Gasoline distributors learned that they could purchase most everything needed for a free-standing service station from HSC. "You could buy the entire filling station—the pumps, the awnings, the building—everything you would

need you could buy from Howard Supply," comments Moody. Despite the brief venture into the gas station supply business, HSC was basically a pipe-valve-fitting distributor for oil companies. As a secondary line of business, the company sold supplies and tools that any oil well service contractor could use.

The period from the mid-1980s to the mid-1990s was a time of boom and bust for the oil industry and this resulted in uncertainty for the firm. The year 1986 was a very bad year in the oil fields, so it was a tough time to be rebuilding a company, because we were literally starting from scratch, Moody explains. "However, while everybody else was losing business in '86 because of the slow down, every dollar we sold was a dollar more than we sold beforehand."

The decade long slowdown in the oilfield resulted in the merger of several oil companies and the disappearance of many independent operators. This trend also affected supply houses and when the two biggest supply companies merged, HSC management knew they had to do things differently if the company was to survive.

It was determined the company could not survive if the focus was not on a particular segment of the oil service industry and HSC set a strategy to divest itself of inventory that was not used by the well servicing contractors. And, from that day in 1995, HSC began to specialize in the well servicing segment of the business. With the decision to become a specialized supplier for the well servicing business, the company went from struggling to survive to a thriving supply company. With that change in strategy, HSC became the largest, most preeminent supply company with a leaning toward specialization in the well service segment that triggered a period of growth and expansion for the company.

Tom Work, who came to the company in 2001 and now serves as technical advisor, remembers that when he came on board, HSC operated seven stores, most of them in the Rockies. Today, HSC has over twenty stores and service operations in California, New Mexico, Texas, Oklahoma, Wyoming, Utah, Colorado, and North Dakota. The company is headquartered in Fort Worth, Texas.

"Our first entry, in California, outside of the standard oilfield supply store offerings was our Wire Rope and Rigging Shop in 2003," Work recalls. "We were approached by a customer who asked if we could help with

Pedro in the tong repair shop.

their drilling line changes, sling fabrication, and other wire rope needs. We had never been involved in this line of business in California; however, our customer was very confident that we would assemble the right people and facility. I knew of one man whose knowledge, reputation, and desire was unmatched in the industry. His name is Steve Lewis. He has been in this line of business since he was in high school, and the only thing more impressive than his knowledge is his work ethic. Together, with the help of a brilliant engineer named Bill Cheney, we developed our state-of-the-art spooling trucks for drilling line and sand line installations. This design remains unmatched in the industry."

Along with a 10,000 square foot fabrication facility being built, a 350 ton horizontal load test machine was added for testing derrick raising lines, slings, chain, and eventually pipe handling tools. "This all took a substantial amount of money to construct,

and bring to fruition," says Work. "At the time we were owned exclusively by the Late family. Dick McMillan was a very fair, and at the same time, very demanding owner. After the planning and presentation to Dick, he pointed his finger at me and said; 'this is my grandchildren's inheritance that I am loaning you, and I don't even have any grandchildren yet!'"

The company growth was also stimulated by several strategic acquisitions. In 2000, HSC acquired E&H Industrial Supply stores in the Rocky Mountains. The acquisition of Justice Supply and Machine allowed the company to move its footprint into Northwest New Mexico. The E&H acquisition brought an immediate presence into three Wyoming locations, and one in North Dakota. The E&H operations have always been very strong in the industry segment of wire rope and rigging, so the fit of the traditional oilfield supply store together with the rigging expertise of the E&H groups was a great addition to

the industry in that area. The acquisition of Justice Supply and Machine in Farmington, New Mexico gave HSC a presence in the Four Corners gas and coal mining market. Here, you can procure supplies for oil, gas, mining, as well as, a complete machine shop in the mold of what a real oilfield machine shop has always been—manual machines to do any task from threading pipe, to a complete draw works overhaul and repair. The manager of the machine shop, Sid Shepard, has worked in that same shop since he was eighteen years old. In his forty-six years, he has seen and worked on every type of equipment imaginable.

Another important acquisition for HSC came in 2005 when the company purchased Wireline Cable Service in Farmington, New Mexico. This was a family owned business that provided an entry into that segment of the market into a brand new area for HSC. The Attaway family had served this market for many years, and Marvin Attaway had

the foresight to install a load test bed capable of testing cable, chain, and other fabricated product long before the needs of today were apparent. This acquisition also brought a cable spooling truck service to serve the Four Corners area.

Above: Another tong repair loaded for the field.

Below: Load test performed on tubing elevator.

Above: Dick McMillan, owner of Howard Supply Company examines a braided sling fabricated by Dustin Moreno (right) during Rope Shop dedication.

Bottom, left: Magnaflux test being performed on a rod elevator trunnion to detect cracks.

Bottom, right: Howard Supply Company hoisting equipment inspection facility.

The product line carried by HSC is much broader than its competitors. The company stocks more than 90,000 items throughout its locations. The ordering and fulfillment of customer orders is handled by the newest and most efficient computers, but old-timers in the company still remember the days of thick catalogs listing thousands of parts and the paper-and-pencil method of writing orders.

HSC has immediate availability of those products a customer needs every day. "We supply the full range of products, but our inventory is stocked according to the geographical markets we serve," explains Work. "For our Rocky Mountain operations, the big product is wire rope. It's their big calling card. Here in California, we carry more than $2 million in inventory just in the Bakersfield location," Work adds. "We always try to have what is needed and stay a step ahead of what our customers are going to need so we can have it when they need it."

One of the most beneficial capabilities HSC has added in

California is an inspection service for hoisting tools. The guidelines are very strict and explicit, as set forth in API Recommended Practice 8B. Processes are followed to the letter regarding the teardown, inspection, NDT (including load test and magnaflux), and documented from the pickup/arrival of the equipment location or at the HSC facility to the delivery of the finished product. This service was pioneered by HSC and its customers in California, and the company is very proud of this service to the industry as a whole.

During 2014 the company had a change in leadership when longtime President and CEO Moody retired after thirty years. Moody was instrumental in positioning the company during the early years and establishing its name as the premiere supply and service company that exists today. Moody was succeeded as president by Michael "Mike" Barber, who came to the company after spending a total of thirty-seven years with Haliburton, and Harbison-Fischer, before joining HSC.

Barber joined HSC in 2012, to work on special projects and supplier management issues. "In a supply company there are always opportunities to improve the supply chain." says Barber. "After many years in executive management with other companies I was excited to be involved in supporting the various operations of the company when the decision was made to change the leadership of the company. It is a great honor to be promoted into such an important position as president and CEO."

Barber feels the biggest contributing factor to HSC's continued success is its commitment to customer service. "We are willing to go the extra mile to help the customers solve their problems, whether it is supply issues or technical repairs. It's that dedication to our customers that has earned our reputation," he says. "If a customer calls, we'll go out after hours or weekends to solve the problem. There is virtually no limit to what we wouldn't do to help a customer out of a bind."

Under Barber's direction, HSC is continuing to grow its presence in the markets it serves. "You can't be all things to all people, so what we've done is continue our focus on the well servicing markets. That puts us in a fairly narrow category relative to the entire oil and gas market. However, it allows us to be specialized in our chosen market space, especially well servicing operations and all the associated products and equipment that go with it," Barber explains.

There are lots of supply stores, but only at HSC will customers find a complete line of products integrated with design, repair and recertification services. From lifting equipment to pipe handling tools, and from specification to recertification, HSC helps its customers extend the service life of their equipment.

"My goal is for our customers to think of us in terms of reliability and trust," asserts Barber. "We want our customers to have confidence that HSC will be a partner to solve their supply chain problems. We want customers to know they can count on us when the chips are down and they truly need a supplier who is dependable and trustworthy. And we want our customers to know that when we make a promise it becomes a commitment. Howard Supply Company stands behind what we do."

Above: Left to right, Steve Wright, Key Energy Services and Tom Work share experiences.

Below: Howard Supply Company pumpjack, California Avenue, Bakersfield, California.

MTS SOLUTIONS

MTS Solutions, which has serviced oil wells in California for thirty-two years, has earned a reputation for the highest standards in service, quality of products and safety. MTS' capabilities include servicing producers, injectors, generators, disposal wells, gas plants, co-generation plants, refineries and line cleanup. The company is headquartered in Bakersfield.

MTS Solutions began as Mitchell-Taylor & Sons in July 1983 when Jay Mitchell and Byrl Taylor purchased the stimulation services division of Champion Chemicals.

Taylor, a native of Bakersfield, began his career with Diversified Chemicals Corp. where he became assistant regional manager for the Los Angeles Basin. He later joined Champion Chemicals and ultimately became western regional manager.

Mitchell began his career in 1974 with Diversified Chemicals. He later joined Champion Chemicals and became district manager in Bakersfield with responsibility for the San Joaquin Valley and Coastal Region Fields.

Even in its early days, finding solutions was what MTS was all about as Taylor and Mitchell found that providing custom design programs produced the best results. "One job in particular, a design program we created, resulted in an increase in production from pre-treatment levels of eight barrels a day to after-treatment production of approximately 100 barrels per day," explained Taylor. "Payout for the individual wells ranged from four days to two weeks. MTS also designed the system to delay plugging the decline curve and showed a dramatic improvement."

Left to right, Monda Byrd, MTS vice president and Tommy Reed, president.

Mitchell, who grew up in Bakersfield and graduated from Bakersfield College, is a third generation oilman. His grandfather, P. D. Mitchell, was a drilling superintendent who pioneered California oil development in the Los Angeles Basin and San Joaquin Valley fields. His father, Jerry Mitchell, was a partner in Western Avenue Properties, the umbrella company for several pump and supply and liquid waste management firms.

Another early assignment involved a water disposal well in a field of North Bakersfield. In the past, the well had been acidized with conventional methods and while the initial response was good, it had proved short-lived.

"With close water analysis, we found a microbiological problem," explained Taylor. "Bacterial slime was plugging perforations and the sand face permeability. The conclusion

was based on observation of samples in waste water treating facilities, which indicated premature plugging of the injection well. Since bacterial slime was essentially impervious to acid, conventional procedures were not successful. After studying the problem, we decided on a different approach.

The two-step program designed by MTS included a pre-soak with a proprietary chemical blend, followed by conventional acidizing.

The solution devised by MTS resulted in an increase from 500 barrels per day to 15,000 barrels per day, which was virtually all the water lease produced. The disposal rate was about thirty-three percent better than when the well was first drilled.

Success stories such as these formed the foundation on which MTS Solutions was built. Today, the company continues to progress with technical advances, the invention of production chemicals to enhance production and assist in well maintenance, and an emphasis on working safely each day.

MTS offers a wide variety of well maintenance products and services. The company also provides traditional chemicals and a line of environmentally friendly green products for those looking for the same effective results.

"Our green initiative is one of the main goals we are working on today," explains Monda Byrd, MTS vice president. "As of last year, we began introducing 'green' chemicals into the marketplace. These chemicals meet the highest standards approved by the EPA and carry the DFE seal of approval. We are working to offer a direct replacement for the harsh traditional chemicals used in the oilfields. These chemicals are highly effective so we anticipate these being part of our new generation of products as we move forward."

The core values at MTS, developed through more than three decades experience, include safety, teamwork, solution and reliability.

At MTS, health and safety is the number one priority. Every member of the team is personally committed—24/7—to his or her own safety and the safety of others, both on and off the job. MTS has a full-time safety manager who oversees the equipment and training needed to keep everybody safe.

The importance that MTS places on teamwork can never be overstated. The MTS 'team' includes both the MTS staff and its customers. The company's commitment to both is absolute and unwavering and 100 percent team satisfaction is the goal.

Because many jobs are unique and require a versatile approach, MTS is focused on finding solutions to often challenging problems. MTS employs the most efficient and cost-effective and laboratory-designed processes to achieve optimal, timely results.

MTS is noted for its reliability because customers know it delivers the results it promises. The company realizes that two things are needed to make a team—leaders and players—and that it takes hard work, effort, leadership, vision and players to make a good team.

"We are committed to meeting or exceeding our customers' high expectations and earning their trust by providing state-of-the-art oil field services, focusing on 'green' product use when possible and always performing with the greatest possible focus on safety, efficiency, real-time execution in complete compliance with regulatory and environmental requirements," explains Byrd.

Taylor retired from the firm in 1993 and Tommy Reed has served as president of MTS Solutions since Mitchell retired in 1989. Reed was operations manager for DI-CHEM Dresser before joining MTS in 1983.

Byrd, who joined the company in 2007, is vice president. Prior to joining MTS, she worked in various accounting and management positions in profit and non-profit organizations.

Don Blurton, the corporate secretary, joined the company in 1983 after working in various management positions for Champion Chemicals and DI-CHEM Dresser.

The management team also includes Gary Starling, who joined the firm in 1988 and is currently manager of Southern California sales, and John Johnston, the plant manager, who joined the company in 1987. Starling worked previously for BJ Titan, DI-CHEM Dresser, Aminoil and proudly served in the U.S. Military. Prior to coming with MTS, Johnston was with DI-CHEM Dresser for seven years. These key individuals are all shareholders of the corporation.

MTS Solutions has forty-three full-time employees, an increase of twenty-seven in the past eight years. MTS team members have decades of experience within the industry, including the experience of well-qualified sales representatives ready to assist with any and all client needs. MTS Pump and Transport staff are trained and experienced to meet all industry standards.

The company also has qualified advisors available to assist customers with their technical and design needs.

MTS is involved with a number of professional organizations, including the California Independent Petroleum Association (CIPA), Western States Petroleum Association (WSPA), and American Petroleum Institute (API). Team members contribute to a wide range of community organizations, including Taft College Foundation, United Breast Foundation, MADD, American Cancer Society, American Farm Bureau, Teen Challenge, Special Olympics, R. M. Pyles Boys Camp, Ronald Reagan Presidential Foundation, Veterans of Foreign Affairs, Children's Wish Foundation, Optimal Hospice Foundation, North High Band, Bakersfield Police Officers Association, Kern County Firefighters Union, and many others.

MTS Solutions offers a variety of products and services necessary for well maintenance operations. The company aims to utilize the best products and methods to maximize production for clients. From traditional products to environmentally-friendly products, the needs of the client will be met by producing beneficial results for all, utilizing the best practices and procedures.

C. D. LYON CONSTRUCTION, INC.

Above: Left to right, Kiki, Chris, Debbie and Jeff Lyon.

Below: Installation of slip line.

In 1985, Chris Lyon left the security of a large, family-owned construction company and decided to strike out on his own. With plans to go into residential construction, Lyon purchased a backhoe and dump truck. However, his former customers in the oil industry kept calling Lyon and asking him to do oil work.

One thing led to another—as they say—and in 1986, Chris and his wife, Debbie, organized C. D. Lyon Construction, Inc., a full-service general engineering contractor capable of handling several types of construction, both in and out of energy-related fields. Nearly thirty years later, Chris is still in the oil business.

Chris' family was involved in oil field construction and he grew up in the business, beginning to learn on-the-job while still in high school. When a 1971 earthquake caused a pipeline break near Northridge, Chris was asked if he could work a few nights to help get it repaired. "The few nights turned into three straight months and I decided to leave school at Ventura College and go to work full-time," he recalls.

Chris served a stint in the U.S. Army, then returned and became a field superintendent before deciding to form his own business.

The company had only two employees at first—Chris and Debbie—and they worked from a dirt lot with a little wooden shed that served as an office. Within two years, the company had grown to sixty employees and its reputation was growing. The company branched out into landfill construction and methane recovery, a decision that led to a major contract with Getty Synthetics, which was using methane to co-generate electricity. This project alone doubled the company's size.

Chris believes the company has achieved success because of its determination to solve problems for its customers. "We have always done what we signed up to do, and we pride ourselves in covering all the bases when we do a job. We don't want to have any call-backs," he says. "We're available 24/7 and if a customer has a problem we go out and repair it."

During summer break from school in Chris' early years, he spent his first day in the industry working in the oil fields by helping to excavate an oil line on a property in Ventura. Twenty-five years later, Chris purchased the very same property and constructed his own company's main office and yard at 380 West Stanley Avenue in Ventura in 1998. The company now has a twelve acre site for storage of its fleet of more than a hundred trucks, as well as an array of Caterpillar construction equipment. The company also owns and operates a facility in Santa Maria.

When C. D. Lyon Construction is engaged for a project, the company takes care of every aspect of the job. The company's skilled and experienced personnel provide customers with engineering, structural concrete and process pipe fabrication, as well as complete

project management. The company takes care of every aspect of the project, including excavation, pipe laying, structural concrete and equipment setting and leveling.

C. D. Lyon specializes in meeting the new construction, maintenance and operations needs of its oil, gas, LNG, petrochemical and chemical customers in both the onshore and offshore sectors. For nearly thirty years, C. D. Lyon has constructed new, and upgraded existing oil and gas plants, installed new and expanded existing reservoir infrastructures, and constructed structural components and/or mechanical systems for offshore platforms and port transfer facilities. The company also offers production transmission services such as pipelines, underground pipelines, pipeline rehabilitation and trenchless pipeline installations.

The company also delivers new construction and maintenance services for power generation and cogeneration plants. Other services include civil and structural component installations, boiler repairs, cleaning and dredging cooling water channels, and process/power piping. During extended plant

Above: Oil dehydration plant, Santa Maria, California.

Below: Produced water plant, Ventura, California.

Above: New gas plant installation, Lompoc, California.

Below: Power plant, Ventura, California.

outages, C. D. Lyon specializes in providing quality, skilled and diversified crews necessary for large scale plant maintenance and expansions such as gas turbine, boiler and coolant system upgrades.

C. D. Lyon provides equipment installation, replacement and upgrade services for new or expanded industrial manufacturing processes. The firm also offers dedicated facility maintenance crews for continuous service of plant components, maximizing process uptime and productivity.

For its solid waste customers, they provide methane gas, storm water, waste water and leachate recovery systems in landfills. Additionally, the company constructs and installs permanent and modular treating facilities for landfills. Some treating facilities burn methane gas to produce electricity as a form of green energy. Others refine the landfill waste fluids, and then blend them with municipal waste water treatment systems. C. D. Lyon also provides pipeline services to transport the methane gas to offsite facilities where it can be used to offset natural gas consumption.

Among the many other services provided by C. D. Lyon are design, build and project management. The company can work as part of a customer's design/construction team, or take a customer's project from concept to budgeting and feasibility studies, through construction and to a successful startup.

The company's well-trained personnel can provide project management services including estimating, scheduling, subcontractor coordination, direct work supervision, design coordination and document control.

As a Department of Transportation certified contractor, C. D. Lyon complies with DOT's environmental health and safety standards and can install new pipelines, or rehabilitate the old, or maintain existing pipelines under DOT supervision.

C. D. Lyon carries an ASME S-Stamp and U-Stamp, authorizing shop and field manufacture and assembly of power boilers, manufacture of pressure vessels, and repairs and/or alternations of boilers and pressure vessels. The company is a fully USL&H licensed and insured contractor, authorized to work on structural and mechanical components of offshore platforms in both California state and federal waters. The company also has both shop and field industrial blasting and coating capabilities.

The firm is proficient in pipeline construction and rehabilitation, process piping and facility maintenance.

With a modern 5,000 foot sand blasting and paint booth adjacent to the company's main lot, the company is able to provide both shop and field industrial blasting and coating capabilities. C. D. Lyon can test, erect, blast and coat various structural steel components ranging from moment frames to underwater high-pressure pipe clamps, crane pedestals, and suspension bridges. C. D. Lyon also excavates places or erects new structural concrete foundations and walls, and demolishes and repairs existing concrete structures.

The highly qualified personnel at C. D. Lyon are available to track a project's progress from material and equipment procurement to completion and assembly of complete Quality Assurance and Quality Control (QAQC) packages tailored to the customer's needs and applicable code requirements.

Environmental health and safety is a vital concern of C. D. Lyon and the company has developed a comprehensive and continuous occupational injury and illness prevention program, which includes DOT-compliant drug and alcohol testing, detailed process management and behavior based safety procedures. The company's experience modifier rating (EMR) has been below industry average consistently since the program was started twenty-five years ago. The company is

firm in its belief that the health and safety of an individual employee takes precedence over all other concerns.

C. D. Lyon employs over 200 people at peak capacity and the company and employees are involved in such community activities as Boys and Girls Clubs of Greater Ventura, local 4-H/FFA junior livestock auctions, API and various local sports teams. The company has a number of long-time employees, including Operations Manager Mike Turk, who has been with the firm since its first year. Dispatcher Vincent Torres has been with the company for twenty years, and Superintendents Larry Elshere, 1992; Joaquin Silva, 1998; and Jeff Danebrock, 2000.

C. D. Lyon Construction, Inc., is in the process of transitioning from Chris and Debbie to their son and daughter, Jeff and Kiki Lyon. Jeff joined the family business after earning a Master's Degree in Civil Engineering from Cal Poly and now serves as construction and engineering manager. Kiki, who also has a Master's Degree in Business from Cal Poly, serves as Controller.

For more information about C. D. Lyon Construction visit www.cdlyon.com.

Above: C. D. Lyon Construction, Inc. specializes in meeting the new construction, maintenance and operations needs of its oil, gas, LNG, petrochemical and chemical customers in offshore sectors.

VICTORY OIL COMPANY

Oil pioneer Edwin P. Crail was born in a small Pennsylvania coal mining town in 1893. As a young teenager, Edwin decided to leave the coal mines and rode the rails to California's Imperial Valley, where he found work planting date palms.

Edwin served as an infantry scout for the U.S. Army during World War I, then returned to California and began working as a roustabout and roughneck in the Lost Hills area of Kern County.

In 1922, Ed Stearns of Universal Consolidated Oil Co. made Edwin "an offer he couldn't refuse." Stearns sold Edwin two hard-tire Packard trucks for an extremely low price, with the condition that the newly formed Crail Bros. Trucking Co. would respond first to any of Universal's calls for service. This relationship endured until the late 1960s when the company ceased operations as a trucking company.

Meanwhile, by 1930, Edwin and his brothers—Laurence, George and Fred—were working together in the oil field hauling business. The company was active in the Taft and Lost Hills area as well as the Los Angeles Basin.

Edwin then relocated the company to Southern California where its trucking company and newly-established crane service company, now located in Long Beach, could service the Los Angeles Basin as well as the central coast and valley oilfields to the north.

The Crail brothers expanded their business to include cranes and established Crail Bros. Crane Service Company, which became one of the region's premier heavy-haul crane and trucking companies. The Crails were pioneers and innovators in the business, and were instrumental in designing a better mobile crane. The company ordered, assembled, operated and serviced mobile cranes.

The entrepreneurial Crail brothers, also started Rotary Materials, Inc., which delivered drilling mud dredged from a local slew to well sites throughout Southern California. The new rotary drills being utilized in the oil and gas industry required drilling mud and the Crail brothers supplied that demand in addition to becoming an oil well drilling contractor. All of these expansions were natural extensions of their core oilfield trucking business and it was only a matter of time before the Crails would form a consolidated company named Crail Bros., Inc., Ltd. and become an oil and gas operator and producer as well.

Victory Oil Company, which had been founded in 1934, was purchased by the Crail brothers in January of 1941. The Crails consolidated all of their oil and gas properties and operations into Victory Oil Company, with Crail Bros. Inc., Ltd. becoming strictly a commercial carriage and crane service company.

During these early years, Crail Bros., and later Victory Oil Company, acquired and developed oil and gas leases throughout Central and Southern California and established production in the Long Beach, Signal Hill, Wilmington and Torrance oilfields of Southern California. Expansion continued up the coast from Ventura County to the Santa Maria and Santa Barbara oilfields, and inland into the San Joaquin Valley from Lost Hills to Cymric to McKittrick and into the vast Midway Sunset oilfield in Kern County.

Edwin was a driven—but fair—businessman who was well-liked and respected in the oil patch as well as in the crane and trucking community. He was also a visionary and invested very wisely. Sam Brown, Edwin's son-in-law, said, "He was the absolute supreme leader. He held this position with such grace that it made his leadership a pleasure to those he led."

Opposite, clockwise starting from top left:

Edwin Crail served as an infantry scout for the U.S. Army during World War I.

The Crail brothers.

Derrick rig up, Long Beach, 1956.

Signal Hill Oilfield.

Clockwise starting from top left:

Lost Hills roughnecks.

The Crail brothers.

Crail Bros. three axle truck and crane.

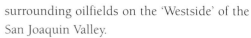
surrounding oilfields on the 'Westside' of the San Joaquin Valley.

During the 1960s and 1970s, Victory Oil Company became a pioneer in the use of steam injection and other thermally-enhanced recovery techniques to develop the substantial shallow heavy oil deposits in the Westside oilfields of Kern County. The heat delivered to the reservoir with the injected steam served to reduce the viscosity of the heavy oil and to increase an individual well's production tenfold or more.

The increased production was welcome good news but the new stimulation technique also presented some significant challenges as well. Oil produced in this manner created very stubborn oil/water emulsions, rendering the production virtually unsalable until a way to break-up the emulsion and remove the produced water could be devised.

Guided by Sam, the company developed high temperature steam generators, heater treaters, free water knock-out units, and a super-heated high pressure gradient vaporiz-er to break-up the toughest emulsions. Aptly named the 'bomb,' the high pressure vaporiz-er actually did blow one afternoon in the mid-1980s.

By 1980 the company's production was in excess of 2,000 BBLs/day and growing rapidly, as its heavy oil reserves were drilled and developed with the infrastructure necessary for thermal recovery operations. With the decontrol of heavy oil prices by President Carter in 1979, San Joaquin heavy crude, which only a few years earlier had sold for $5-$6/BBL, was fetching a far more attractive $25/BBL in 1980. With these market changes, the value of heavy oil properties increased dramatically, as witnessed by Shell Oil Company's $3.6 billion acquisition of a neighboring independent heavy oil producer, Belridge Oil Company, in 1979.

Riding this trend, in 1984 Victory Oil Company sold its larger heavy oil properties to a group of major companies, including Sun Oil Company, Shell Oil Company, and Union Oil Company. Shortly thereafter, Victory sold its producing properties in the Los Angeles Basin, while continuing to operate its remaining smaller properties in

Victory Oil Company maintained offices in Long Beach and Taft, California, since the 1930s and 1940s, and the corporate offices were recently relocated to San Pedro, California.

Edwin passed away in June of 1977, while driving from his home in Palos Verdes, California, to the Kern County oilfields which he had done so much to develop.

Edwin's son-in-laws, Robert 'Bob' Johnson and Sam J. Brown—who married Edwin's twin daughters Jerry and Patricia Crail—decided to try their hands at the oil business in 1956, with the purchase of the McKittrick Front lease from local oilman Jack Goforth. Jerry and Sam were geologists and Bob studied petroleum engineering and the law.

The first generational transition was now under way and by the 1960s, Bob and Sam were running Victory Oil Company, Crail Bros. Crane Service Company and Crail Bros. Trucking Company operations, under the guidance of Edwin. Bob managed the Long Beach office where the crane and trucking business was located and operated the Los Angeles Basin oil and gas properties. Sam ran the Taft field office where the company had accumulated extensive heavy oil reserves in the giant Midway Sunset and

Kern County. During the 1980s, the next generation of Johnsons and Browns took up the mantle of independent oil producers and continued to operate Victory from a new Taft field office.

By 1986 the oil price pendulum had reversed course and heavy oil was again selling in the $6/BBL range. In 1987, Victory took advantage of the depressed commodity price and acquired heavy oil properties at an attractive price from Sun Oil Company (then Orix Oil & Gas). In 1988 the Johnson family acquired the Brown family's interest in Victory Oil Company. Bob Johnson's son, Eric, became president of Victory Oil

Company in 1989 after holding positions ranging from roustabout, pumper, geologist and operations manager. He continues as president today. Along with his brothers Alan and Craig and his sister Ann, the Johnsons remain active in the operation of Victory Oil Company.

Following the path blazed by previous generations, Victory Oil Company sold its remaining operated heavy oil and gas properties in 2012 to various larger oil companies, including Vintage Petroleum, AERA, and Seneca Resources, among others, at a time when San Joaquin heavy oil was selling for more than $100/BBL.

Opposite, top: Neely lease jack line.

Opposite, middle: Robert Johnson during World War II.

Opposite, bottom: Robert and Jerry Johnson.

Above: Victory Oil Company steam generator.

Below: Oil field tour, 1970s.

BREITBURN ENERGY

Hal Washburn and Randy Breitenbach, close friends since college, met in Randy's apartment in Los Angeles in 1988 to put the finishing touches on their lifelong dream of starting their own oil company. "We had no money, no business plan," Randy admits. But the two twenty-seven year olds were convinced that they could take advantage of rapidly-evolving technological advancements in the oil industry to increase production from mature fields and do it on their own.

Breitburn—a play on Hal and Randy's surnames—was born.

Starting with two wells in California, Breitburn Energy has successfully implemented its founders' strategy of acquiring long-lived oil and gas fields, raising production with state-of-the-art technology and experienced operating teams, and increasing cash flow from those assets. Today, less than three decades later, Breitburn is among the country's largest and most successful independent oil and gas production companies.

Hal and Randy first met in the early 1980s at Stanford University where they both graduated as petroleum engineers. Soon thereafter they became roommates; ultimately, they became lifelong friends.

After graduation, Hal moved to Denver and began drilling wells with a regional operator and Randy moved to Alaska as a facilities engineer before joining Arco's finance group

in Los Angeles. Despite their different initial paths, their entrepreneurial spirit burned brightly and the two spoke often of their idea to form their own oil company. After a few years of discussions, the two men were ready to proceed. Hal moved from Denver to Los Angeles, and in early 1988, the two rented a house in Hermosa Beach that doubled as their first office.

At the time, the oil industry was experiencing one of its periodic downturns, and the major oil companies were leaving the lower forty-eight and focusing their efforts offshore and overseas. To fund their new ventures, many of these companies began selling their U.S. assets, and one of the first places they exited was California. Hal and Randy were convinced that California was a massive opportunity, home to some of the largest producing oil fields in the country with billions of barrels of oil in place and low-decline rates. With backing from family and friends, Hal and Randy bought two wells in Ventura County and the Alamitos lease in the Seal Beach Field.

The two friends ran a lean operation. After the Hermosa Beach "office," they worked from

Clockwise, starting from the top right:

Alamitos lease, May 1990.

Alamitos sunrise.

Left to right, Hal Washburn and Randy Breitenbach, hard at work in the field.

an eight foot trailer at Seal Beach, doing whatever was required: signing checks, stamping envelopes and driving forklifts and backhoes. They routinely worked sixteen hour days to keep their dream alive. With production doubling on the Alamitos lease and steady performance from the Ventura wells, the new venture was on its way.

Their timing could not have been better. The new company was able to hire from a deep pool of skilled geologists and engineers that had become available as the major oil companies had turned their attention away from California. At the same time, desktop computers were becoming powerful enough that small companies like Breitburn could use the same tools as larger oil companies to perform sophisticated geologic and reservoir modeling.

After a few years of steady growth, Hal and Randy learned of an opportunity that would ultimately transform their business: Occidental Petroleum was selling its interests in the West Pico and Sawtelle Oil Fields in Los Angeles. The Sawtelle Field is located under the Veterans Hospital in West Los Angeles, and the West Pico unit of the East Beverly Hills Oil Field is located in West Los Angeles and Beverly Hills, both some of the most expensive neighborhoods in the country. Because of the heightened sensitivities of operating in such urban areas, only a handful of firms was interested in the properties and Breitburn was the successful bidder, closing the deal in 1993.

Breitburn's plan to modernize the West Pico site to produce previously untapped oil reserves was opposed by some residents who were concerned about the potential effect that a growing production facility might have on their neighborhood— despite the fact that oil had been produced from the site since the mid-1960s. After years of public hearings, protests and lawsuits, Breitburn's plans were finally approved; construction began in 2000, and the improvements resulted in a significant reduction in noise and cleaner air. Today, visitors to the neighborhood are surprised to learn that the West Pico facility, which sits behind decorative walls bordered by manicured landscaping and whose lone drilling rig is enclosed within what easily passes for a clock tower, is home to an oil production facility.

Despite the initial challenges, the Sawtelle and West Pico Fields fit the founders' strategy of buying mature oil fields and using tools such as geologic modeling and reservoir simulation to optimize the assets. Since being acquired by Hal and Randy, the West Pico and Sawtelle Fields have produced more than 12 million barrels of oil; ultimately, they are expected to produce more than 22 million barrels, almost four times more than the 6 million barrels of proved reserves that were in place at the time of the acquisition.

As the 1990s came to a close, Breitburn had become a stable, successful venture but needed additional capital to continue growing. After securing $30 million of private equity funding, Hal and Randy made another large acquisition in the Los Angeles Basin: they purchased the Santa Fe Springs Field, Rosecrans Field, Brea Olinda Field and a portion of the East Coyote Field from Texaco in 1999, effectively tripling the size of the company. Once again, the timing

Clockwise, starting from the top:

Room with a view: A rig in the Sawtelle Oil Field in West Los Angelos, Memorial Day, 1995.

The lone drilling rig at the West Pico production facility easily passes for a clock tower.

The Seal Beach Field.

Top, left to right: Santa Fe Springs Field, then (1929) and now (recent times).

Below: Randy Breitenbach at the first office in May 1990.

Bottom: The East Coyote Field.

was right: Hal and Randy took advantage of a low commodity price environment to conclude the deal, and they then acquired the rest of the East Coyote Field from Unocal in 2000 during the same low price environment. These fields were a perfect fit for Breitburn's strategy: they were large, complex accumulations of oil and gas whose value could be enhanced by applying modern technology.

The Santa Fe Springs Field, a giant oil field located approximately fifteen miles southeast of downtown Los Angeles, is a good example. Discovered in 1919 and covering approximately 1,100 acres, Santa Fe Springs has approximately 2 billion barrels of original oil in place and at its peak produced 345,000 barrels of

oil/day. To date, more than 2,000 wells have been drilled in Santa Fe Springs and cumulative production stands at approximately 643 million barrels. When Breitburn acquired Santa Fe Springs, the field was producing 1,400 barrels of oil/day; more than fifteen years later, at the end of 2014, Breitburn was producing 2,700 barrels of oil/day—nearly double the production rate when it acquired the field. In addition, Santa Fe Springs has produced approximately 10.5 million barrels of oil since being acquired by Hal and Randy; ultimately, Breitburn expects to produce almost four times the 5.8 million barrels of reserves in place at the time of the acquisition. Today, Breitburn sees even more potential at Santa Fe Springs and plans to expand its production handling facilities to accommodate its future growth plans.

In addition, the Texaco acquisition provided Breitburn with additional upside as the deal included a significant real estate component. In a collaborative effort involving Breitburn, the city, neighbors and developers, the land was sold, remediated and rezoned for commercial and residential use. Today, several commercial buildings have been built and more than 550 homes and residential units are planned for The Villages at Heritage Springs, a residential community in Santa Fe Springs, on land that was part of the original deal.

Breitburn's continued growth created additional needs. After outgrowing the Seal Beach trailer and a couple interim offices, in 2000, the growing Breitburn team established their corporate headquarters in downtown Los Angeles—ironically, they moved into the same office space previously occupied by Arco, where Randy started his finance career. In 2004, Hal and Randy raised approximately $140 million in exchange for a piece of their growing business, allowing them to continue to implement their development strategy by hiring additional

geologists and engineers and expanding their use of technology.

In October 2004, Breitburn acquired the Orcutt Hill Oil Field in northern Santa Barbara County. The field, which has been producing since 1901, is named for William Warren Orcutt—incidentally, another Stanford engineer. The self-educated geologist is credited with overseeing development of the field in the early 1900s, which ultimately led to the creation of four oil towns in California: Bicknell, Graciosa, Orcutt and Orcutt Hill. Today, the only surviving community is Orcutt, a suburb of Santa Maria.

Breitburn had grown from start-up into a larger organization and the next step was to tap the public markets. On October 10, 2006, Hal and Randy successfully took Breitburn Energy Partners LP public as a master limited partnership, nearly two decades after they had acquired their first two wells in California—at the time, the company had an enterprise value of approximately $400 million. Breitburn's common and preferred units trade on the Nasdaq exchange under the symbols BBEP and BBEPP.

At the time of the IPO, Breitburn Energy Partners LP consisted mainly of the California assets that Hal and Randy had acquired over the years in addition to properties in the Wind River and Big Horn Basins in Wyoming. Breitburn quickly expanded in 2007, adding properties in the Permian Basin, Florida, and the Midwest, primarily Michigan.

Under Hal's leadership—Randy retired in 2012 but remains involved as vice chairman of Breitburn's Board—Breitburn continued its successful acquisition and development strategy. Between 2011 and 2013, Breitburn acquired more than $2.1 billion of oil and gas properties, including a position in California's prolific Belridge Field in Kern County for approximately $94 million. In November 2014, Breitburn completed the largest transaction in its history when it acquired QR Energy, LP for approximately $2.5 billion. After closing the deal, Breitburn was the largest oil-weighted upstream MLP in the United States with estimated proved reserves of 315 million barrels of oil equivalent, an enterprise value of approximately $6.7 billion and more than 900 full-time

employees across its operations, including nearly 200 in California. At the end of 2014, Breitburn was producing approximately 57,500 barrels of oil equivalent/day, almost thirteen times more than it was producing at the time of its IPO less than ten years earlier.

When recently celebrating Breitburn's twenty-fifth anniversary in business, Hal reflected, "I am extremely proud of what Randy and I started, but we didn't do it alone. Breitburn would not be the company that it is today without all the fine people who have contributed over the years and who still contribute each and every day. While we have expanded our portfolio over the years to seven different producing areas, including Ark-La-Tex, the Permian Basin, and the Mid-Continent, we'll never forget that it all started in California. With a proven history of success, Breitburn is well-positioned to successfully execute its acquisition and development business plan for decades to come."

Above: Left to right, Hal Washburn and Randy Breitenbach celebrate Breitburn's twenty-fifth anniversary at the Nasdaq Exchange.

Below: CEO Hal Washburn.

There he built a beautiful home with a view of the distant San Gabriel Mountains. Shortly after moving in, an air raid siren wailed as unidentified planes were spotted near Long Beach. World War II was underway and there were fears that the local harbors would be bombed. This proved a false alarm, but Jack installed a bomb shelter behind the home, just in case.

Jack was saddened in 1949 when his trusted friend and partner, Paul, died of injuries received in an auto accident. After Paul's death, Jack partnered with oil man Daniel Elliott to form Wilmington Oil Well Development Company and drilled a lease near the harbor area.

To take advantage of continuing drilling activity, Jack organized the International Drilling company, with his nephew, Ben Herley, as manager. International contracted drilling jobs with various production companies. A new oil field had developed along a Southern Pacific Railroad track near Oxnard and Jack obtained a lease from

the railroad for a one mile, 100 foot wide right-of-way through the oil field. As the track ran in the center of the strip, only fifty feet was available to install the drilling equipment. It was a tight fit, but three wells were drilled that produced heavy, thick, six gravity asphalt.

In the 1950s, Jack began suffering from a strange malady that baffled his doctors. However, he was well enough to attend the wedding of his son, Jim, to Shirley Madison, a beautiful and talented girl from South Dakota. Jack's new daughter-in-law suggested that he visit the famed Mayo Clinic in Rochester, Minnesota. Doctors at Mayo recommended surgery for a possible brain tumor. However, the problem was actually an aneurysm, for which no diagnosis was available at the time. Jack agreed to the operation but survived only a few hours after surgery. Jack was only fifty-eight years old. At the funeral, St. Barnabas Church was filled with oil men and civic leaders, as well as family and friends.

Jack's son, Jim, a musician with a degree from the University of Southern California, assumed the management of the company after Jack's death. Jack had insisted that Jim complete courses in accounting and oil field technology, and that knowledge proved invaluable. Jim partnered with the son of another oil man, Charles Cather, and together they bought leases containing three oil wells.

The direction of the business changed in 1960 when the 405 freeway cut a swath through Long Beach north of Signal Hill. Herley Pipe and Supply Company was in the path of the new freeway and was sold to the State of California. On the recommendation of real estate appraiser T. F. Merrick,

Clockwise, starting from the top:

Marlin fishing, Catalina Island.

Home in Long Beach.

Jim Herley at the office in 1960.

A new oil field had developed along a train track near Oxnard and Jack Herley obtained a lease from the railroad for a one mile, 100 foot wide right-of-way through the oil field.

the proceeds of the sale were invested in three acres on the east side of Signal Hill and three concrete tilt-up buildings were built for industrial rentals. This proved to be a successful venture and several other properties were added as the business continued to prosper.

The family's remaining oil wells and real estate properties are now managed by Jack's grandson, David, a 'chip off the old block.' David is an avid hunter and fisherman like his grandfather. He earned a degree in business administration and learned the business working in the oil fields and being mentored by company superintendents. Jack's granddaughters are also successful in their careers. Erin is a certified psychotherapist, and Myralyn is a farmer and real estate developer.

At the time of his untimely death, Jack was the operator of more than thirty oil wells in eight Southern California oil fields. He lived a life filled with adventure and achievement and earned the respect of all who knew him. His family and associates honor his memory and are grateful for his legacy.

Above: Jim and Shirley Herley's wedding.

Below: The Herley Family, 2010.

NAFTEX

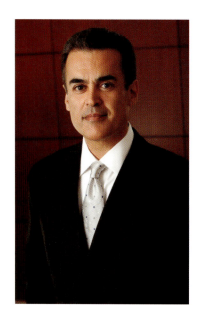

Clockwise, starting from above:

Hormoz Ameri, CEO of Naftex.

Original vintage 1988 caricature of Hormoz Ameri drawn on the back of the send-off card signed by Hormoz's colleagues at Bechtel Petroleum, Elk Hills Oilfield. Hormoz resigned from Bechtel in 1988 to start Naftex at the age of twenty-eight.

Naftex senior staff. Left to right, Daniel Newsom, superintendent; Geoffrey Nicholson, senior staff geologist; Hormoz Ameri, CEO; Charlie McKay, area manager; Anthony Bonfanti, controller; and David Lefler, senior staff engineer.

In 1988, Hormoz Ameri, a young petroleum engineer working for Bechtel Petroleum in the oil fields around Bakersfield, noticed that the major oil companies were beginning to sell off some of their non-core assets. Although the oil industry was in one of its periodic slumps, Hormoz saw an opportunity to establish his own oil and gas exploration and production company.

Hormoz had begun his career as a reservoir engineer with Bechtel Petroleum in 1985 after earning his Bachelor of Science degree in Mechanical Engineering from California Polytechnic University and a Master's degree in Petroleum Engineering from Stanford University. The job provided him with valuable experience in heavy and light oil production, water flood projects, dry gas zone projects and interacting with management. All was going well until around 1986 when the industry drifted into a severe down cycle.

"That's when I noticed the major companies were selling off marginal properties and trying to focus on their core assets," Hormoz explains. "My wife and I were a young couple and didn't have any kids, didn't have a mortgage or any major liabilities. I had always wanted to do something on my own, a sort of an entrepreneurial bent, and it seemed a good time to jump out there and do my own thing. I figured that if I failed, I had done a good

job for Bechtel and they would want me back. So, a down time for the industry proved to be a positive time for me."

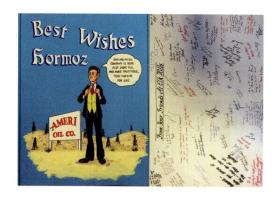

Hormoz, aided and supported by his wife, Fariba, founded Naftex Holdings in partnership with his older brother, Jamsheed 'Jim' Ameri, a real estate entrepreneur. The two brothers had no capital of their own to purchase oil assets for their own account. The alternative they came up with was to raise a modest amount of funds by selling limited partnership interests to friends and family with Naftex Holdings acting as the general partner of the partnerships. Naftex Holdings would purchase the assets for the partnership and charge management fees for operating the properties and earn an equity position upon meeting certain economic returns to its limited partners.

Naftex Holdings started off with a well-developed business plan that emphasized four major principles: acquire long-lived assets with low-risk exploration and development opportunities in mature producing energy basins; optimize reserve recovery by using technical expertise and state-of-the-art technologies, including 3-D seismic, cutting edge drilling technologies and reservoir simulation modeling; reduce cash flow volatility through commodity price and interest rate derivatives and hedging; and maximize asset value through operating and technical expertise.

"The idea was to make some decent acquisitions and make them work by cutting costs and doing away with the many layers of expenses involved in the operations by the major companies," Hormoz says.

The newly formed management company made its first acquisition in 1988 by purchasing a small four well lease in the Mountainview Field in Kern County, California. Before long, Hormoz was able to hire his first employee and Naftex Holdings was off and running.

Naftex made a 'breakthrough' acquisition from Chevron in 1991, when the company obtained the Racetrack Hills heavy oil assets in the Edison Field in Kern County. The Chevron acquisition enabled the company to hire additional employees and begin a period of positive growth. Shortly after this acquisition, Jim decided to focus his full attention on his thriving real estate business in the Pacific Northwest while remaining a shareholder of Naftex Holdings (now Naftex Operating Company), which is the position he holds to this day.

Naftex Holdings grew and became more profitable as a property management company over the next few years. But Hormoz was always searching for attractive opportunities to augment his property management operations by founding and growing companies with direct equity ownership of oil and gas assets. A major opportunity came in 1998 when a severe industry downturn had the big oil companies once again willing to sell key assets at attractive prices.

"It was the right time to make a big move," Hormoz explains. "So, in 1998, my wife, Fariba and I put everything on the line, even our personal savings, formed a new company, which we named Naftex ARM, and negotiated a major acquisition from Chevron. The acquisition included over 4,000 acres of fee land with associated oil and gas production in three major oilfields of Midway-Sunset, Asphalto and Railroad Gap, all located in Kern County.

"1998 proved to be the breakthrough year for me," Hormoz says. "I was able to hire a professional staff, field operators, mechanics and other personnel and to purchase a bunch of equipment."

Under Hormoz's leadership, the Naftex group of companies has consistently ranked among the top twenty oil and gas production companies in California over the past decade. Naftex's consistent long-standing business strategy is highlighted by the drilling of more than a hundred oil wells and cumulative production of more than six million barrels of oil equivalent from properties exclusively acquired from major integrated oil companies

Left: Left to right, Howard Coleman, legal counsel to Naftex since its founding, and Hormoz Ameri, CEO of Naftex.

Right: A group of Naftex field personnel at a company barbecue event at Naftex's Edison Oilfield.

Above: Hormoz Ameri, CEO of Naftex Operating Company (center) on the front cover of The American Oil & Gas Reporter Magazine *as chairman of California Independent Petroleum Association (CIPA).*

Below: The Ameri family on their usual summer holiday in the south of France. Left to right, Hormoz, Roxanna, Katia and Fariba.

such as Chevron USA and Mobil Oil, while selling oil to major refineries such as Shell and ConocoPhillips.

A strong believer in the importance of industry trade associations, Hormoz is a long-time member and supporter of the California Independent Petroleum Association (CIPA), a nonprofit, nonpartisan trade association representing approximately 500 independent crude oil and natural gas producers, royalty owners and service and supply companies. CIPA members represent approximately seventy percent of California's total oil production and ninety percent of California's national gas production. CIPA is an advocate of free market

principles, eliminating duplicative regulation, stimulating recovery of domestic resources and improving the industry's public image.

Hormoz is the past chairman and member of the executive committee of CIPA and has represented the diverse interests of its membership before the California State Legislature, the United States Congress and numerous federal, state and local regulatory agencies.

In recognition of his many contributions to the association, Hormoz was awarded the 2012 Glen Ferguson Award, conferred on an individual who has greatly contributed to the success and advancement of California's oil and gas industry.

Hormoz is also the past chairman and emeritus member of Stanford University's Petroleum Investments Committee (PIC) and two-term member of the Board of Advisors of Stanford's School of Earth Sciences. The PIC oversees and provides direct investment advice for the management of the Petroleum Investments Funds (PIF), which provide an annual income stream for use at the discretion of the Dean to achieve the strategic objectives of the School of Earth, Energy & Environmental Sciences and to further its academic mission. The PIF is the school's largest source of flexible revenue to further its academic mission. PIF principal, with a current market value of $54 million, is invested in producing oil and gas royalties and other energy-related assets.

The Hormoz and Fariba Ameri Graduate Education Endowment Fund in Earth Sciences at Stanford's Graduate School of Earth Sciences was established in 2007 to support talented graduate students pursuing their interests in the areas of multiphase flow, reservoir engineering, reservoir stimulation and thermodynamics. The impetus for the establishment of this fellowship is rooted in Stanford's generous support and the funding of Hormoz's entire graduate education at the School of Earth Sciences in the mid-1980s.

The Hormoz Ameri Energy Systems Laboratory at the California Polytechnic University's School of Engineering was established by Hormoz in 2001, exactly twenty years after he received his Bachelors of Science degree from the school. The learning

environment provided in the Energy Systems Laboratory is designed to meet the demands of changing technologies and industrial methods, providing students with a state-of-the-art, hands-on, polytechnic education. Hormoz, along with Flour Corporation and General Dynamics Corporation, was among the earliest contributors to the construction of an additional 250,000 square feet of laboratory and instructional space at Cal Poly.

"I'm a product of the public schools, so as soon as I was able I gave back to Cal Poly where I did my undergraduate work," he notes. "We established the graduate education endowment at Stanford since Stanford paid for my graduate studies and actually gave me pocket money to live on. Without their full support I would not have been able to go to school and complete my graduate education."

Proud of his heritage, Hormoz serves as a Trustee of the Farhang Foundation, a philanthropic organization with a mission of promoting Iranian art and culture to the community at large. Under his leadership as Chairman of Farhang's Nowruz Committee, an annual event which celebrates the Iranian New Year has grown into the most recognizable and well received event of its kind within the Iranian community and the public at large in Southern California. Hormoz also leads Farhang's Generations Council, which engages the younger generation of professionals who are proud and passionate about their Iranian heritage and wish to contribute their talents and resources to promoting

Farhang's core mission. Hormoz is also a key member of Farhang's Strategy Committee and has contributed to many of the groups cultural, educational and artistic pursuits.

In 2002, Hormoz took advantage of an opportunity to divest some of Naftex's properties but the company continues to grow and remain active in the industry. The staff now includes around thirty-five employees and the firm produces about 1,000 barrels of oil per day. Naftex is headquartered in Los Angeles and operates a field office in Bakersfield.

Fariba has played a key role in the growth of Naftex over the years. "She has been very supportive from the very beginning, and without her, none of this would have been possible," Hormoz says. "She gave me the freedom to take the risks involved in building the Naftex group of companies." Hormoz and Fariba are blessed with two daughters, Katia and Roxanna, both of whom attend Stanford.

Hormoz is also grateful for the support of such long-term employees as Dan Newsom, who was the very first employee hired by the company. Newsom, who recently retired after twenty-five years of exemplary service, started as a field operator and worked his way up to become the company superintendent of operations.

Despite the cyclical nature of the oil industry and the current down cycle, Hormoz remains confident about the future of the oil and gas business. He is optimistic that the industry can anticipate a steady climb in oil prices from 2017 onward.

Hormoz Ameri and his two daughters Katia and Roxanna on a drilling rig.

MACPHERSON OIL COMPANY (MOC)
MACPHERSON ENERGY COMPANY (MEC)

Macpherson Energy Company, an energy company headquartered in Santa Monica, is one of California's leading family-owned oil producers. The company also owns renewable energy projects associated with its oil production activities. The firm's operating companies include Macpherson Oil Company, Macpherson Power Company and Macpherson Green Power.

Macpherson Oil Company is currently the eighth largest producer of onshore oil in California, with principal operations in the Round Mountain Oil Field, northeast of Bakersfield. The company also has operations in the Sharkstooth Oil Field, Midway Sunset Oil Field and Mount Poso West Area Oil Field in Kern County.

Macpherson was founded in 1981 but the company roots go back to 1959 when Donald R. Macpherson, Sr., left Pauley Petroleum to become an independent oil producer. His first oil development was in the Citronelle Oil Field in Alabama. In 1973, Macpherson and his son, Donald R. Macpherson, Jr.,—operating as National Petroleum Associates—purchased a series of leases in the Round Mountain Oil Field.

Oil had been discovered in Kern County in 1899 and the Round Mountain Oil Field discovery well was drilled by the Elbe Oil Land Development Company in 1927. The Round Mountain field developed quickly, with production peaking in 1938 at more than 10,000 barrels per day. However, production declined steadily after World War II and by the 1960s, production was less than 2,000 barrels per day.

In 1981 the Macphersons organized Macpherson Oil Company and began development of the north portion of the Round Mountain Oil Field. Five years later, Macpherson secured the south portion of the oil field from Shell Oil and in 1992, the company secured the center portion from Texaco. Other Macpherson operations were launched at the Mount Poso West Area, and Midway Sunset Oil Fields elsewhere in Kern County.

Macpherson is now a leading independent oil producer, utilizing its expertise in geology, drilling, recovery technology, pumping design, and oil field operations to maximize production.

Through its operations, partnerships and joint ventures, Macpherson has more than 300 onshore wells producing California's oil. Most of the wells are located at the 2,000 acre Round Mountain Unit in the heart of the 6,000 acre Round Mountain Oil Field.

Macpherson brought in Round Mountain's first horizontal drilling well in January 2009. There now are more than 170 horizontal wells, which have increased the oil field's production five-fold.

Macpherson Oil Company has expertise with enhanced recovery operations such as water flooding and steam flooding, making it possible to extract more oil from fields than can be produced through primary recovery operations. Macpherson has implemented secondary and tertiary oil recovery operations including water flooding and steam injection, and has also utilized and perfected horizontal drilling techniques in the fields it operates. Implementation of these new technologies has substantially increased recovery rates and oil production for Macpherson.

In addition to its oil production and operations, Macpherson is fifty percent owner of the Mount Poso Cogeneration Plant, which runs on 100 percent wood waste biofuel and produces forty-four megawatts of electricity. The plant takes in water used in the oil recovery operation and converts it to steam in a boiler using renewable fuel. It then uses the steam to drive a turbine to generate electricity and provide steam for a steam flood of the oil field. Macpherson has been a partner since the Mount Poso Cogeneration Plant opened in 1989 using clean coal technology. In 2011 the plant was converted to renewable biomass fuel.

Renewable energy from Mount Poso is sold to Pacific Gas and Electric Company under a long-term power purchase agreement that helps PG&E and the State of California reach their renewable energy targets.

A corporate affiliate of Macpherson Energy Corporation, Macpherson Green Power, was formed to focus on—and hold—its interests in this important power plant project. Macpherson's partner in the Mount Poso Cogeneration Company is DTE Energy Services, which has biomass projects elsewhere in California and the United States.

Macpherson is also exploring other opportunities for clean energy development.

The Macpherson companies continue to be family owned and operated. Macpherson Energy Corporation holds the majority of ownership in Macpherson Oil Company and Macpherson Power Company. Corporate headquarters are in Santa Monica, and principal operations are at the Round Mountain field, near Bakersfield. Other operations are elsewhere in Kern County as well as Alabama.

Macpherson has seventy employees and depending on the pace of development provides 150 to 200 additional full-time jobs for employees of on-site contractors. Macpherson's experienced and diverse employees include petroleum engineers, geologists, drilling technicians, operators, safety and maintenance specialists and financial experts as well as administrative personnel.

Macpherson Energy has created and maintained a safe and sound work environment for its employees and contractors—meeting or exceeding all safety and environmental regulations. Oil industry activities are measured closely, and Macpherson's safety record for its employees is outstanding.

In particular, Macpherson employees at the Round Mountain Unit reached an important milestone in 2015 when they passed the four year mark without a recordable injury. The safety culture at Macpherson engages all employees at every level of the company, making certain every individual goes home safe and sound at the end of their workday.

Macpherson is a family-owned business where people are valued and have an opportunity to grow. At Round Mountain, a barbecue is held each month for every person in the field. Owners, employees, vendors and contractors meet and share the camaraderie of the company's safe and sound operations and celebrate a job well done.

With its creative culture, Macpherson continues to explore new technologies to create new solutions. At the same time, the company values experience. Many of Macpherson's oil reservoir engineers, geologists, production engineers, field workers, and on-site supervisors have more than twenty years of hands-on experience.

To learn more about Macpherson Oil Company/Macpherson Energy Corporation, please visit www.macphersonenergy.com on the Internet.

THE JEWETT FAMILY

The first family of California oil claims a rich heritage that dates from the very earliest days of the nation. Centuries before Solomon Jewett discovered oil bubbling from the ground on his famous Rio Bravo Ranch in 1896, members of the Jewett family had already distinguished themselves as early pioneers, innovative farmers, political leaders and astute business men and women.

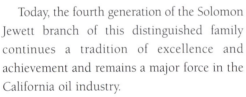

breeding Merino sheep, and soon became a wealthy farmer. One of his sons, Solomon Wright Jewett, Sr., born in 1838, eventually became one of the most prominent men in Vermont. Besides being one of the foremost farmers of Addison County, he was active in state politics and came within one vote of being nominated for governor. Secretary of State Daniel Webster sent Jewett to Europe to convey dispatches from the U.S. Government to the Court of St. James and the Government of France. He served as a member of the Vermont legislature in 1838-1839.

Clockwise, starting from the left:

The Jewett Coat of Arms.

Solomon Wright Jewett, Sr., 1876.

Solomon Wright Jewett, Jr., and Philo Dennis Jewett, 1880.

Today, the fourth generation of the Solomon Jewett branch of this distinguished family continues a tradition of excellence and achievement and remains a major force in the California oil industry.

The Jewett family roots go deep into British history and members of the family were among the very earliest to colonize what would eventually become the United States.

The Pilgrims landed at Plymouth Rock, north of Cape Cod, on November 9, 1620, and established a colony in what is today's Massachusetts. The system of government established by the brave pioneers became the bedrock for the ideals that would lead to establishment of the United States, eighteen years later. Joseph Jewett, son of Edward Jewett of Lovingshire, England, came to America on the ship *John* in 1638 and settled in Rowle, Massachusetts.

One of the descendants, Samuel Jewett, settled near Weybridge, Vermont, began

By the late 1830s, Jewett, Sr., was one of the biggest sheep farmers in Vermont, and certainly one of the wealthiest. He organized the State Agricultural Society, started the State Fair and was sent to London by President Millard Fillmore to represent American agriculture at the Great Exposition of 1851, a world's fair at the famed Crystal Palace in London. During that visit, Jewett purchased sheep from Prince Albert, the husband of Queen Victoria.

Among Jewett, Sr.'s ten children were two sons—Solomon Wright Jewett, Jr., and Philo Dennis Jewett—who appeared to have inherited the family's spirit of adventure. As Katharine Jewett Tierney, Jewett, Sr.'s great-granddaughter, tells the story, Jewett, Sr., gave the two sons $50,000 each to seek their fortunes, although little suggested that the brothers would prosper except for their indomitable spirits, hunger for adventure, and their lofty goals.

"They traveled through the south and considered buying a big plantation, but were deterred by the high cost of purchasing and maintaining slaves," explains Tierney. "Next, they visited Texas but found land there too expensive. Hearing about the gold rush in California, Solomon and Philo decided to head there."

California was still the Wild West in the 1800s with little civilization and constant battles with the Indians. The great California gold rush had peaked but, undeterred, the brothers started looking for land and traveled north as far as San Francisco. They found there were not large enough land holdings and traveled further, soon learning that a number of Spanish land grants were still available at good prices in Kern County. The brothers investigated, saw the possibilities, and used their funds to purchase 120,000 acres. Much of the land was purchased for fifty cents an acre, although the rich bottom land along the Kern River cost as much as one dollar an acre. A portion of the land along the river became the prosperous Rancho Rio Bravo. The brothers acquired surface land and all water rights to the Kern River and mineral rights.

Drawing on the knowledge and experience they had gained by working on their father's sheep farm in Vermont, Solomon and Philo decided to try their hand at raising sheep in California. The brothers invested in huge flocks of Merino sheep. To shepherd the flocks, Basque sheepherders were recruited from the Pyrenees Mountain Range along the border of France and Spain. A perilous ocean voyage on Clipper ships carried the sheepherders with the herds of sheep from Spain, around Cape Horn, to California.

To improve the sheep herd and avoid too much inbreeding, boatloads of rams were imported from the Pyrenees each year. The sheep would arrive in San Francisco and were then driven like a herd of cattle all the way to Kern County.

The sheep produced huge quantities of wool, much of which was sold to the government to make uniforms for soldiers during the Civil War. The brothers made a lot of money selling wool and Rancho Rio Bravo became a big operation.

Providing wool for the Union Army during the war did not set well with those supporting the Confederate cause and, following the war, Philo narrowly avoided an assassination attempt by the Bushwhackers, a group of violent Confederate sympathizers who wished for California to secede from the Union and become an independent country.

Solomon Wright Jewett, Jr.

Thinking their plan would work if they got rid of all the Union supporters, the outlaws vowed to kill all the men who had supported the Union during the war. One afternoon, two of the Bushwhackers headed for the Jewett ranch. They asked for some food and one of the ranch employees, a Mr. Johnson, fed them before realizing why the men were really there. The outlaws killed Johnson but Philo managed to hide until the Bushwhackers left.

Solomon Wright Jewett, Jr., and the Ming family in the Winton Six Touring car, February 18, 1910.

In 1867, swollen waters from the Kern River swept away all the original buildings at the Rio Bravo Ranch. In response, the brothers rebuilt the structures on higher ground. A newspaper editor who was entertained at the ranch following the flood wrote that he was "handsomely entertained" and wondered at the "charming mountain scenery, fine growth of vegetation, and the style of the accomplished eastern farmers."

The discovery of oil on the Jewett lands was a serendipitous event. In 1896 the brothers were drilling water wells on their property to provide watering stations for the sheep herds. "When one of the wells was drilled, they got this black stuff instead of water," explains Tierney. "They knew what the black stuff was, but at that time there was no earthly use for it. They built a fence around the well to keep the sheep out and forgot about it."

Meanwhile, the population began to grow as others were attracted by the opportunities available in the San Joaquin Valley. Solomon became one of the founding fathers of the city

of Bakersfield and soon became one of the town's most successful businessmen, building the first store in Bakersfield at Nineteenth Street and Chester.

At the time, there were no banks between Los Angeles and Stockton and Solomon thought a bank would help the new city grow. In 1874 he opened the first bank in the city, the Kern Valley Bank. When the entire city of Bakersfield burned to the ground in 1889 it was Solomon and his bank that provided the money for businesses to rebuild and begin again.

In 1899, Solomon sold his sheep and turned to raising cattle but before too long the growth of the automobile industry created a demand for gasoline and the well that was drilled for water—but produced oil instead—became the foundation for a highly successful oil business.

Solomon's son, Solomon Wright Jewett, brought the first automobile to Bakersfield, a 1910 Winton Six Touring car. Soon, growing automobile traffic created a demand for better roads and Solomon decided to mix gravel with some of the oil flowing from his wells to produce an early form of asphalt. He then became responsible for paving most of the roads in the area.

Solomon Wright Jewett, followed his father in the oil business and did 'wildcatting' with his good friend, George Getty, Jean Paul Getty's father. Solomon Wright Jewett's son, Philo Landon Jewett, worked with his father in the oil fields.

His grandson, Philo L. Jewett, enjoyed working in the oil fields and was overseeing the drilling in Section 19 of the family's Round Mountain Field in 1936 when drillers hit a gusher that produced 14,000 barrels a day. Philo rushed home to tell his father, who was bedridden with a bleeding ulcer, about the gusher. When he gave him the news, Solomon, Jr., whispered, "That's just fine, son." And then he closed his eyes and died.

Responsibility for the family business fell to Philo but he detested the paperwork and administrative duties that went along with running a large business. His love was engineering and he much preferred to be out in the fields working.

Fortunately, Philo's daughter, Katharine Jewett Tierney, had a 'head for business' and became an accountant after graduating from UCLA in 1960. In her early career, Tierney worked for a business management firm that handled motion picture stars and professional athletes, including. Hall of Fame baseball pitchers Sandy Koufax and Don Drysdale. It was Koufax who introduced her to her husband, Don Tierney, a pitcher for the Chicago White Sox.

Tired of the administrative burdens, Philo turned the company operation over to Tierney in 1971, and she has guided the company successfully since then, representing her two brothers and their five cousins.

In 1992, Macpherson Oil Company consolidated several leases, including the Jewett one, two and three leases in the Round Mountain Field, securing a partnership so that necessary funds could be raised to build a multimillion dollar steam injection plant, to start a secondary recovery program which has been a great success.

Macpherson Oil is currently the eighth largest producer of onshore oil in California, with its principal operations on the old Jewett Oil Field on Round Mountain.

The partnership also operates three large tank farms. All are filled to near capacity, waiting for the price of oil to rebound.

"The oil business is a very risky business," comments Tierney. "It's tough, hard work but our family has managed to be successful in the business for more than a century. However, I advise—to remain in the oil business—one must be diversified in investments, or one can go broke at any time.

The remaining family member carrying the Jewett name is Tierney's nephew, Philo Wright Jewett. He is the only son of Tierney's brother, Philo Lawrence Jewett.

Above: Left to right, Katherine M. Grey, Nancy M. Ditzler, Katharine Jewett Tierney and Philo Lawrence Jewett standing next to the Jewett Family plaque, 1971. The Beale Memorial Clock Tower is in the background. The clock tower is adjacent to the Kern County Museum in Bakersfield.

Left: The plaque was placed by the Kern County Historical Society and the Kern County Museum in 1970.

PHOTOGRAPH COURTESY OF DAPHNE FLETCHER.

CAVINS OIL WELL TOOLS

Since 1928, well servicing and drilling operations have depended on high-quality, American-made Cavins Oil Well Tools and replacement parts to keep them up and running—safely and dependably. Among the company's best known—and most utilized—products is the 'Advance' Spider, the pioneer in power slips and still the leader.

Until the 1950s, the Signal Hill headquartered company was engaged primarily in manufacturing, renting and selling proprietary oil well clean-out tools such as sand line hydrostatic bailers and pumps used by oil and gas operators to service their wells.

In 1951, Waldo Moore brought the original design for the 'Advance' Spider to Cavins and the company became the sole manufacturer, patent owner and distributor of the product line, which soon became popular throughout the oil fields.

Harry Dawson, a mechanical and aeronautical engineer who had served as a major in the Royal Canadian Air Force during World War II, went to work for Cavins as the manufacturing manager in 1961. Harry's son, Jim, recalls that there were plenty of oil fields near his father's home in Long Beach, but it was not until he grew tired of spending hours on the road each day that he began looking for a job closer to home. Harry applied at the Cavins Company for the position of shop manager and was quickly hired by the company founder, Paul.

In 1963, Harry, along with an uncle and another investor, purchased the company and incorporated it as Cavins Oil Well Tools. Two years later, the firm expanded by purchasing the Tork-Master division of General Controls. Harry had worked for that company before joining Cavins and was familiar with its primary product—an automated valve actuator.

In 1973, Harry bought out his uncle and the other shareholders through a stock redemption arrangement. Later that year, the

Right: Original patent drawing for Cavins bailer, filed in 1930.

Bottom, left: Signal Hill discovery well.

Bottom, right: Signal Hill Oilfield.

Opposite, top left: Oilfield workers are using the Cavins 'Advance' Spider, widely used since 1950.

Opposite, top right: Well servicing crew with Cavins bailer and junkbasket.

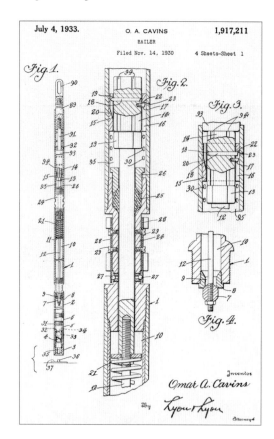

The company was founded by Paul and Omar Cavins to rent their patented oil well tools, which included hydrostatic bailers, sand pumps and junk snatchers.

Tork-Master division was sold to help facilitate the payoff of the former shareholders and the remaining debt was refinanced through a bank.

In 1974, Cavins acquired Taft Welding & Machine Co., which later became a Cavins Repair Center in Kern County.

Harry went into semi-retirement in the late 1980s and turned over the day-to-day operations of the company to his son, Jim Dawson.

Jim started working for the company part-time when he was only nine years old and believes that working his way up the ladder provided a great way to truly understand how a company operates at a deeper level.

Jim recalls that his first job assignment was compiling loose-leaf catalogs for customers and doing odd jobs around the office during holidays and summer vacations. "Dad always said, 'If it's not done right, it's half-baked.' He instilled many qualities in me, including being responsible and always striving for excellence."

After attending UC-Irvine where he majored in chemistry, Jim joined the family business full-time. He started in the advertising department and later worked in purchasing, managed the rental tool division, managed sales and marketing, and then became executive vice president and chief operating officer.

Jim promoted and negotiated the major acquisition of the well servicing product lines of Varco International in 1997. This added power tongs, hooks, links, elevators and other items, including licenses for the widely known and respected brands of Baash-Ross, Foster, Hillman-Kelley, Kelco, Varco BJ, and Web-Wilson. The acquisition of Varco virtually quadrupled the size of the company's product lines and almost instantly doubled sales volume.

In a 2012 profile published in *Well Servicing* magazine, Dawson said, "A valuable lesson I took away from this process was the importance of integrity in building and maintaining relationships. Once the negotiations were completed, we were informed by the president of Vacro/BJ that what really consummated the deal was the mutual trust we had for each other."

Right: Cavins 'Advance' Spider.

The magazine article went on to describe Jim as a hands-on CEO who prefers to get out of the office. "The most enjoyable part of my job is working with the best people in the industry toward our common mission of providing superior quality products and services that meet our customer's requirements and exceed their expectations," he said. "Our loyal and devoted employees represent the most important part of our organization."

Surprisingly, Jim still insists on signing the company checks—an old-school tradition established by his father.

Dawson told the magazine writer that much of his time is spent in meetings to stay informed and keep his finger on the pulse of the business. His biggest challenge is finding enough hours in the day to accomplish the many operational goals the company is trying to meet, while dealing with increasingly severe government regulations, taxes and control. "The cyclical nature of the industry is another challenge for our strategic planning," he said.

Cavins Oil Well Tools has always been known for quality, initially with a variety of OEM down-hole clean-out tools and, later, with well servicing pipe handling equipment. Perhaps the company's best know product is the Cavins 'Advance' Automatic Spider for casing-tubing and drill pipe, first introduced over sixty-five years ago.

Years of experience and testing—and many improvements—have made the Cavins 'Advance' Spider the leader for safe, economical, and trouble-free handling of tubular goods. Heat treated alloy steel is used throughout for the greatest possible strength and the longest wearing life with a minimum of weight. The shafts are hardened alloy steel for longer life. Replaceable steel bushings are fitted into the journal bearings and all bearings and journals are equipped with grease fittings for lubrication. Air cylinders and foot valves are mist lubricated through the airstream. All Cavins 'Advance' Spiders are equipped with a manually operated safety latch, which may be used to positively lock the slips in their set position. Because of the

low operating pressure, the slips cannot release from the tubing until it has been raised by the elevators, thus guarding against a lost string in the event that the foot valve is tripped accidently. The slips, shafts and air operating mechanisms are protected by guards to prevent damage caused by the elevators or other outside forces. Through research and field testing, Cavins engineers are constantly looking for additional improvements to assure the safest and most maintenance free equipment available.

Jim sees the ever-increasing "blind acceptance" of counterfeit tools and replacement parts as one of the most important issues confronting his business. He feels the counterfeit tools compromise the long-standing tradition and commitment within the industry of using the highest-quality products.

A driving force in the use of counterfeit parts is their lower cost, but Jim feels this factor could be overcome if companies would take the time to thoroughly evaluate or run tests on their own to compare tools or parts. Using counterfeit parts is also a dangerous practice, especially when an end-user decides—perhaps unknowingly—to use counterfeit replacement parts and a failure occurs. "In most instances, they have just accepted full responsibility for the incident, which will likely have an adverse effect on claims experience and future insurance costs," Jim comments.

Jim feels the best way to avoid counterfeit products is to look for the OEM trademark identification on the products or insisting that certificates of authenticity be provided with each order.

"We have spent countless hours perfecting our products and testing them against competitors with inferior heat treating and looser tolerances, many with substantial or 100 percent foreign content," Jim says. "The cheapest does not always mean the most economical, and we feel it is of paramount importance in these difficult economic times that we keep our jobs in the U.S. and not export them. That's why our products contain only domestic raw materials and are manufactured by Americans. We couldn't be prouder to label our tools, "Born in the USA®.""

Cavins is headquartered at 2853 Cherry Avenue in Signal Hill. This is the same location where offices were destroyed during a famous 1940s incident when an oil tanker lost its brakes on Signal Hill. The company also has offices in Bakersfield, Taft and Ventura.

The company has several hundred active customers. The top ten customers account for less than twenty percent of the firm's annual revenue, with no single customer totaling more than about ten percent. Export sales comprise about forty-five percent of Cavins total volume.

Cavins began with only five employees in 1928; that number has grown to approximately sixty-five today. The company and its employees support a number of civic activities, including contributions to the Oil Workers Monument in Taft, West Kern Oil Museum and the Long Beach chapter of the Salvation Army. The company has also participated in the Westec oil workers training school and contributes to national and local scholarship funds sponsored by the Association of Energy Service Contractors.

Looking to the future, Jim is cautiously optimistic that the oil and gas industry will continue to stabilize. This, however, is tempered by a continued skilled labor shortage and increased governmental regulation fostering more foreign competition.

For Cavins Oil Well Tools, the business plan for the future is to maintain industry leadership in providing the safest, highest quality oil well tools, and continue to grow through additional acquisitions and in-house innovation.

Cavins original office was demolished by an oil tanker that lost its brakes on Signal Hill, which is located above the office. The building was rebuilt and still exists at the same location.

Dole Enterprises, Inc.

dba AKA DB Pump Supply; DD Natural Resources, LLC; Gordon Dole Company

A Tribute to Gordon 'Tom' Dole 1939-2013

Gordon 'Tom' Dole checking on Crimson Well #1.

Gordon 'Tom' Dole is fondly remembered by family and friends for his incredible work ethic, dedication to his business and concern for others. With sheer determination and hard work, Dole built a successful crude oil production company based on old school morals, family values, and honest, hard work.

Gordon was in his early forties and had been working as a truck driver foreman when he stumbled on the opportunity that allowed him to fulfill his dream of acquiring his own oil wells. In the early years, he acquired his first lease that had no production, but did have the potential to produce again if he could locate a gin pull. Gordon

and his former wife, Arlene, drove an old Lincoln Town Car and carried oil field tools in the trunk of the car. Joe Rose remembers, "What a sight to see...the back end squatted and the shocks about to give out."

As Gordon's long-time friend Joe tells the story, Gordon knew he could get the lease producing if he could locate a gin pull he could afford. Joe had a rig that had been abandoned on one of his leases and Gordon asked Joe if he could have it for his lease.

"The old rig was an ancient single pole, single drum rig mounted on an old truck that someone had adapted from an old Ford flat-head V-8 engine, 1932 vintage," Joe recalls. "There were no doors on the rig and no hood to cover the motor compartment. The rig had been stripped of all its tools and a small sapling about ten feet high had grown up through the motor area."

Despite the age and condition of the old rig, Gordon felt he could use it to pull shallow wells and Joe told him he could have the rig if he could get it off his property.

"A day or so later, my son, Kent, told me that Gordon had the old engine apart and had a big wrench on the crankshaft and a hammer hitting the wrench trying to free up the crankshaft. Of course, everything was covered in oil," Joe says.

"About two days later, Gordon called and asked if I wanted to watch him drive the rig from my lease back to Bakersfield. I was busy, so I missed the event, but my son witnesses the old rig going down Woody Highway at an estimated two-and-a-half miles an hour, with Gordon standing on the running board with one hand on the steering wheel and one hand holding a small container of gasoline dribbling gas into the carburetor to keep the engine running."

Gordon drove the 1910-1920 vintage rig back to Bakersfield and used it to do well work on some of his early leases until he was able to acquire modern rigs. The original gin pull that helped start Dole Enterprises is still displayed in the backyard of the Dole home in Bakersfield.

Gordon drilled his first oil well in 2012 and named it KaPeg (Crimson #1). "KaPeg was his baby and he checked on it every

day while it was being drilled and was involved in everything every step of the way. It was his proudest accomplishment," notes his wife, Karen Dole.

A number of key individuals have been involved with Dole Enterprises from its early days to the present. Karen worked with her husband from the beginning of their relationship and now runs the company. Longtime friends and advisors included Joe Rose, Mike D. Lindeman, Tom Ladd, Gary Taylor Wimberly, Fred G. Rappleye, Ron Anderson, Mel and Richie McGowan, and Bill Alexander. There are too many to mention that helped Gordon along the way.

The crew, which was like family to Gordon, included David Church (employed 19 plus years); Shanon Webb (employed 15 plus years); Richard Jenkins (employed 7 plus years); and Mark Sanders (employed 6 plus years). "They all referred to him simply as 'dad' because he helped raise most of them and because that was exactly what he was like to them," explains Karen. "He treated his crew like family at all times. He taught, listened and helped them from work to home life as much as he could. The crew remains an important part of Dole Enterprises and is helping the company succeed by running the field operations."

Jacque Pittman, a Dole Enterprises secretary and close friend for more than thirty years, also played an important role in the company's growth. She recalls that when the company first started and before there was any crew or enough funds to hire workers, it was just Gordon. "There were times when he would be out in the field working so hard he'd come back to the office covered head to toe in oil and dirt. He would go straight to the bathroom to clean up and I'd have to follow right behind because what was on him ended up on the walls and all over the sink," she recalls.

Gordon Dole covered in dirt and oil from working in the field.

Gordon had a knack and intuition for fixing things and had an amazing work ethic and dedication. He spent his entire life working long hours, seven days a week,

Gordon and Karen Dole.

with few or no breaks. While other small businesses often collapse under pressure, Gordon succeeded by keeping his business in-house and doing as much of his own work as possible to reduce costs. Even during time when oil prices were good, he stuck by the old saying, "I'm saving for a rainy day." Family and friends explain that he remained humble and never lived above his needs. "He stuck to the basics and believed that through hard work you could achieve it all," commented one friend.

Gordon checked his oil wells daily to make sure they were running properly and he recruited others to do 'drive-by checks.' His daughters Bobi and Tiffany, learned to check on well's and gauge tanks at a young age and got most of their practice for their driver's license test by driving in the fields with their dad checking on oil wells.

Gordon's family and crew spent many days in the field with him, eating dinner on the job and helping when they could. Gordon worked so many hours that his wife would bring coffee to the rig so he could stay awake. Karen worked alongside her husband many days, nights and weekends, gauging tanks, fixing equipment, and handing him tools. Gordon and Karen referred

to these activities as their 'dates.' His family understood how important the business was to him, so they spent numerous days in the field, just to spend some quality time with Gordon.

Tiffany remembers a day when she got to the office after a business meeting. She noticed her father was not in the office and although she was in nice dress clothes, she decided to take some coffee out to the lease. "When I got there, my dad was fixing the drive line on the rig," she recalls. "My dad was extremely tall, so he was lying underneath the rig. As soon as I got out of the car I heard him yell, 'Hey, come over here. I need you.'

"Well, you never tell your dad 'No', so despite my strongest intuition about the fate of my nice clothes, I crawled under the rig. Dad was too tall to kneel under it, and his arms weren't quite long enough to reach up at the correct angle, so he asked me to help him loosen a bolt. Of course, I was covered with oil from the drive line and my nice clothes were ruined, but I can still hear my dad laughing and asking, 'Those weren't your good clothes were they? Stop and get some Dawn dish soap on your way home.'

"Come to think of it, I don't think anyone who lived in our house had a piece of clothing that didn't have a dirt or oil spot on it. It was our life and livelihood, so it was a normal thing for us," she adds.

Although Gordon loved his business and spent most of his time working, he did find time for hobbies. He had a pilot's license and loved to fly to Laughlin and Pismo Beach. He was an animal lover, particularly dogs, and one of the organizations close to his heart was H.A.L.T (Helping Animals Live Tomorrow) Rescue. He admired eagles and collected art featuring the majestic birds. He also enjoyed taking vacations on cruises where he could relax and get away from the phone for a few days.

Family members are proud of the fact that Gordon helped the people around him as much as he could. "You could walk the streets of Bakersfield and hear numerous people testify to him giving them not a hand-out, but a hand-up," says Karen. "He invested in rental properties with the profits from his oil wells and gave many people a chance to have a home and even forgave many months of late rent. He was willing to help anyone who wanted to help themselves and there were times when he would buy a vehicle and let people make monthly interest-free payments so they would have a way to get to and from work. Gordon never forgot the people who helped him get his start, and wanted to pay it forward as much as possible."

Gordon died on June 28, 2013, and Karen, his widow and best friend, now serves as president of Dole Enterprises, Inc. The company remains a small, 'family' business with a four man crew that has been with the company from 6 to 19 years, and a secretary who has been with the firm more than 30 years.

"We hope to keep Gordon's legacy alive and continue to do what Gordon loved—oil," says Karen. "He is truly missed in everything we do and we often resort in our critical thinking to asking, 'What would dad do?'

"Gordon was Dole Enterprises and we couldn't explain anything about the company without fully expressing the man behind it all. We want to thank everyone who supported us during the difficult days after Gordon's passing. We wouldn't have been able to keep his dream and legacy alive without the numerous friends, business colleagues, and family. We hope Gordon is smiling down on us, knowing that he 'Taught us right.'"

Gordon Dole.

ENERGY TUBULARS, INC.

B&B PIPE AND TOOL CO.

Energy Tubulars, Inc. (ETI) was founded by Bob Braly in 1959 to distribute oil country tubular goods. Today, ETI is a full-service provider of world-class tubular goods solutions, providing cost-effective solutions from supplier to well site delivery and beyond.

B&B Pipe and Tool Co. has serviced the petroleum, geothermal and waterwell industries since 1952 and has grown into one of the largest pipe slotting service shops worldwide. The company was established in 1926 and was purchased by the Braly family in 1950. It is currently owned by Craig E. Braly, the grandson of C. E. Braly.

C. E. Braly started in the supply end of the California oil industry in 1920 in Taft. He was transferred to Long Beach in 1927 when the drilling boom was started by the Signal Hill discovery and subsequent discoveries in other Southern California locations.

Shortly after the Braly family purchased B&B Pipe and Tool, the 1950 earthquake shook Long Beach Harbor, shearing off many onshore wells and creating a need for major pipe repair. The disaster proved a lucky break for B&B. Another major upheaval, the Korean War, created a pipe shortage in 1952, requiring round-the-clock pipe threading and additional opportunities for the company.

In 1963, Energy Tubulars was appointed as authorized distributor of Oil Country Tubular Goods (OCTG) for NKK of Japan,

a major, world-class manufacturer of OCTG. The company's growth was further aided in 1965 when oil drilling began on THUMS oil islands in Long Beach Harbor.

ETI has become a full-service OGTG solutions provider to oil, natural gas and geothermal customers throughout the industry. Through strong partnerships with mills, strategically placed distribution facilities, and quality tubular service, ETI completes the OCTG supply chain.

ETI's oil and gas customers realize improved costs via its centralized management of the OCTG supply process. Optimizing OCTG inventory is what ETI does best. Its staff works closely with drilling companies to establish optimum inventory quantities while maintaining reduced material costs. Strong partnerships with mills such as US Steel and Tenaris enhance ETI's ability to provide quality value and savings.

Right: The THUMS oil island in Long Beach Harbor.

ETI's logistics services and storage facilities enhance its OCTG supply chain. Reliable rail transloading, truck transportation, online inventory reports and digital POD's with electronic signatures are among the many solutions provided by ETI. In addition, inspection, threading and reclamation services round out ETI's commitment to full-service ECTG solutions. ETI reduces customer's exposure to ownership of inventory and associated management costs.

Growing through the years, ETI now supplies customers throughout the United States, with a concentration in California, Colorado, Wyoming, North Dakota, Ohio and West Texas.

These strategically located facilities from the Pacific Coast through the rocky Mountain region to a Midcontinent hub in Tulsa enables ETI to service oil and natural gas companies for their OCTG requirements in key drilling areas. ETI is also a long established tubular goods supplier to geothermal exploration, development and production companies.

As an added-value service, ETI operates a wholly owned subsidiary, Tubular Transport and Logistics (TTL), for tubular goods storage, online inventory management and trucking to well sites. TTL currently has facilities in Rifle, Colorado; Bismarck, North Dakota; and Bakersfield.

Strong OCTG partnerships develop through cooperation and commitment. Through ETI's many years of experience in the industry it has developed solid, long-standing relationships at every level of the OCTG supply chain.

ETI strongly values all of its business partners and recognizes that strong, reliable relationships have been key to the company's success over its many years in the oil country tubular supply business. ETI works with leaders in their fields to ensure the delivery of quality products and valuable services to its customers.

B&B Pipe and Tool Company has serviced the petroleum, geothermal and waterwell industries for more than sixty years. The company has grown into one of the largest pipe slotting service shops worldwide. From two key locations in California—Long Beach and Bakersfield—B&B provides unexcelled, prompt service to customers worldwide.

With its extensive operating equipment and highly skilled operators, B&B Pipe and Tool provides the highest standards of workmanship with twenty-four hour service to meet customer deadlines for product quality and on-time delivery.

The experienced personnel from B&B are always ready to discuss its customer's needs and help them determine the most effective individual service.

Petros Tubular Services, a sister company, provides third-party tubular inspection and tubing reclamation services, as well as trucking services, to California oil companies.

The Braly family has been involved with many charitable endeavors in Long Beach during the past forty years. Long Beach Memorial Miller Children's Hospital lobby is named the Mary Alice Braly Reception Center. The hospital's Todd Cancer Institute reception center is named for Robert and Mary Alice Braly.

The Bralys are also involved with many other charities including Ronald McDonald House, Long Beach Day Nursery, Assistance League, Ability First, Memorial Hospital Women's League and Stramski Center.

The plan for the future is to continue the legacy of these family-owned companies in the supply and servicing of oil country tubular goods in California and wherever other opportunities may present themselves.

PHIL RYALL, PETROLEUM GEOLOGIST

In July of 2015, Phil turned eighty-one and, after spending fifty-five years as a petroleum geologist, you would think that he had been ready to retire. Well he was not! He had many projects still in mind, both in petroleum projects and social issues. What bothered him the most was the misunderstanding of climate issues (change or warming). He believed there was no proof of man causing climate change or that there was any extreme change. He was hopeful that his extensive experience would have been used as a advisor for energy companies or attorneys in energy related cases.

His career started in 1960 after he graduated from Fresno State with a degree in geology. The only job he found was mud logging for Core Laboratories. After nine months of living around the country in different basins logging wells, and having a new wife and baby, he applied to Shell Oil Company in Bakersfield as a geological tech, which morphed to subsurface geologist for Shell in California Exploration and a very good career with a world class oil company.

For almost ten years, Phil enjoyed the company of—and training from—Shell Oil. He had many interesting projects in California Basin petroleum studies, but wanted to work overseas. So he quit Shell in 1970 (his father and father-in-law thought he was fired). Then he struck out on his own to become a World Wide Petroleum Basin Expert (that did not work out).

In 1980, Phil partnered with Guy Burge, an ex-Union Oil geophysicist, and they put together an exploration program in California with investors from the U.S. and Canada under the name of Stockdale Energy Company. After leasing some prospects for oil and gas, the world oil price dropped. Burge left and Phil struggled to get some of the prospects evaluated—hard times!

Phil Ryall.

During the last drilling episode and the loss of a potential large gas discovery he joined with a landowner/investor, Pardee Erdman. He and some people Phil hired started the company, Stockdale Oil and Gas, with Phil as president/manager in 1987. The company picked up several bargain leases due to the low oil prices and started some redevelopment and extensions. The upshot of this was a company with about 2,000 MCF of gas and 1,500 BOPD. After seven years, the Erdman family decided to sell the company to Oxy, much to Phil's disappointment.

So then Phil was back in the consulting business with several prospects in his hip pocket and hoping for oil and gas prices to recover soon!

The one great thing Phil had was a family of three children, fourteen grandchildren and a lot of good memories. And many good friends in the oil industry and out!

WESTSTAR

In little more than twenty-seven years, Weststar has grown from a small trucking company to become the preeminent pipe transportation and logistics solutions provider in the industry.

The company was founded by Dan Corriea in October 1988 with only three trucks and nine trailers. Initially, the company operated from a ten by fifteen foot office on a two acre tract in Fontana, California, where the trucks and trailers were parked. Dan's wife, Margie, handled invoices and other administrative duties at home on the kitchen table. Dan performed all the minor maintenance on the equipment himself. Other key individuals in the firm's early days were operations and yard manager Jack Jennings and dispatcher Howard Stute.

In its early days, Weststar's primary business was hauling pipe from the Los Angeles harbor and CSI Steel Mill in Fontana to various locations in California. After about a year in business, Weststar rented a larger, ten acre yard in Fontana from Kaiser Steel and, in 1990, Chevron persuaded Dan to be its tubular transport company and move its equipment to Bakersfield.

Also in 1990, Jim Varner became a financial partner in Weststar, with Dan continuing as the managing partner.

During the decade of the 1990s, Weststar purchased two hydrocranes and additional equipment. As the company grew, successful relationships were forged with such firms as Shell, Santa Fe Energy, Elk Roofing, and Occidental. In 1997, Mobil and Shell formed Aera Energy and Weststar began a centralized logistics partnership with the new firm.

Weststar's success, to a great degree, resulted from the firm's commitment to modern technology. "We are more data and technology driven than anyone else and we review the most current data to anticipate our customer's needs," explains Dan. "We use new technology all the time to manage communications and coordinate material deliveries." He adds that Weststar was using this technology before others were even talking about it.

Weststar established a separate software company, Raptor, to develop the sophisticated software needed to stay on top of rapidly changing technology. Weststar also leads the industry in incorporating global positioning systems (GPS) fleet wide, an innovation that allows dispatchers to track deliveries quickly and accurately.

"We have reinvented ourselves a lot," Dan says. "We're always looking for better ways and continuous improvement with logistics."

By October 1997, Weststar operation had grown to 3 hydro-cranes, 10 trucks and 40 trailers. This growth prompted the company to purchase a larger office building and shop at 5760 East Lerdo Highway in Shafter, California.

During the early 1990s, Weststar had focused on drilling and supporting operations, but in 1998 the company began to diversify into production and well maintenance operations. In partnership with Aera, processes were developed to successfully support well intervention activities. 'One Stop Trucking,' a centralized concept was developed in 2000 and Weststar was awarded Aera's total logistics contract.

Also in partnership with Aera, a material flow concept to support production operations was introduced in the Belridge Field in 2001.

In February 2006, Dan Correia purchased a local company and created Atlas Crane & Rigging, Inc., a heavy haul and large crane venture.

As Weststar's centralized logistics concept continued to grow, the company was approached by Occidental in 2010 to manage a call center for its Permian Basin operations. Two years later, Hess Corporation signed an agreement with Weststar to provide data for their fluid and completion operations in North Dakota.

Weststar's load count peaked at 49,000 in calendar year 2012.

Dan feels Weststar has been successful because it leads the industry in new concepts. "We have different ways of handling material that nobody else has, and we do things nobody else is doing," he notes.

Weststar can also be characterized by its commitment to innovation. The company changed how tubular and rods are delivered to well serving rigs in the early 2000s by designing and deploying Material Flow Trailers for Aera Energy. Again—with the Casing Hook System (CHS) in 2009—Weststar revolutionized how OCTG tubulars are offloaded at drilling sites by mitigating more than eleven hazards the industry had just come to accept as normal. Seeing how everyday work is done from a different perspective to make Weststar's employees safer and provide efficiencies to its customers have maintained Weststar's position as the leader in oilfield logistics innovation.

Weststar is also an industry leader in safety and has developed effective programs to protect both its employees and customers. Weststar was the first trucking company to receive the Cal-OSHA Golden Gate Award and the first to achieve Sharp Certification with Cal-OSHA (Voluntary Protection Program.) The company has also achieved a satisfactory rating with the California Highway Patrol BIT program, with eight consecutive audits passed through 2015.

Weststar has grown from six employees in 1990 to 140 at the end of 2014. An estimated twenty-five percent of the company employees have been with the company ten years or

more. The company and its employees are involved in a number of charitable activities, including Links for Life, BCHS, Pyles Boys Camp, Relay for Life and the CBVI Center for the blind and visually impaired.

Looking to the future, Weststar is developing a Green Fleet strategy for California and expanding truck and crane operations in Texas. Weststar is focused on growing its yard management and Inventory Solutions Division, expanding its Logistics Control Center business and to continuing to develop innovations in employee safety.

For more information about Weststar, visit www.truckingweststar.com.

ATLAS CRANE & RIGGING, INC.

Atlas Crane & Rigging, Inc., a large crane and over dimensional hauling company, was established in 2006 when the owners of Weststar realized there was a niche market for such services in Kern County.

Although a free-standing entity, Atlas Crane & Rigging is located next door to the Weststar headquarters on East Lerdo Highway in Shafter. Atlas shares the same innovative spirit and dedication to meeting the needs of its customers that has made Weststar an industry leader.

Since Atlas was established, the company's services have expanded from Kern County into many other counties, cities and states.

Atlas specializes in large and small crane work with transportation of over-dimensional loads. The company can supply hydraulic cranes from 19 ton to 90 ton, along with three-axle tractors geared for handling large sized loads while being compliant with the California Air Resources Board guidelines.

Atlas also has an assortment of trailers that can transport anything that needs to be moved. The fleet includes highbed trailers as well as a 9- and 13-axle, 5-axle lowboy, drop deck, double drop deck, stretch, and split-axle trailers that can handle loads in excess of 150,000 pounds. Atlas also operates a fourteen acre storage yard with an 18,000 square foot building for enclosed warehousing.

The ownership of Atlas boasts nearly fifty years' experience handling, hauling and storing tubulars and equipment and prides itself on its superior customer service. Dispatchers are available twenty-four hours a day, seven days a week, and a unique GPS tracking system allows the company to know where its assets are at any time. Atlas has also equipped its vehicles with cameras to provide 360 degree vision around the trucks, trailers and cranes to eliminate blind spots while maneuvering around tight locations.

Atlas' number one priority is its commitment to safety and keeping all of its employees safe. The company has recorded more than 1,800 days without a recordable injury. This enviable record is due to the company's safety programs and the incredible employees who participate and believe in the programs. Atlas is also involved in the OSHA Voluntary Protection Program, and was the second trucking company in California to achieve the status of Golden Gate Certified.

Atlas has forty well-trained employees who guarantee that your job is done in an efficient, organized and timely manner. Atlas Crane & Rigging is a Minority Business Entity registered with the State of California.

The values that guide Atlas are well summarized in the company's Statement of Values: "At Atlas Crane & Rigging, Inc., we place value on honest relationships with our employees and customers and encourage creativity, quality and respect."

For more information visit www.atlascrane-inc.com.

CIMARRON OIL, LLC

Above: Tony Rausin at Mon #24.

Below: Tony Rausin at the Mon Lease Stock Tanks.

Cimarron Oil, LLC, a small independent oil producer in Bakersfield, California, is owned and operated by Anthony "Tony" Rausin. The company was founded in 1987.

"It all started when my parents moved to Bakersfield from Oklahoma in 1952," Tony explains. "Shortly after arriving in Bakersfield, Dad went to work for Clyde and Iris Smith, who owned the Bald Eagle Oil Company. The Bald Eagle Oil Company owned a small fee property in the heart of the Kern River Oilfield. Mr. and Mrs. Smith lived on the property and Dad and Mom had a small house nearby.

"When I was born in 1958, the oilfields were my first home. The property had about twenty wells, which were all operated from a central jack plant, which was common in the early years of California oilfields. We lived there until 1967 when the property was sold to Tenneco Oil Co. Those first nine years on the Bald Eagle were the happiest and most formative years of my life, in a unique and special time and place. Dad went to work for Tenneco Oil Co. as a pumper, and I rode along with him as much as I possibly could on weekends and during the summer. About age fifteen, I had seen enough and decided that I wanted to have a small oil company. That was my goal from then on."

In 1985, Tony went to work for Tretolite as a sales engineer and spent nine years treating oil and water on the east side of the San Joaquin Valley. During that time, he started pumping a small lease in the Mount Poso Field for Stockdale Oil & Gas. In 1995, Stockdale sold the Mon lease to Tony and some partners and they started Cimarron Oil LLC. For the last twenty years, the Mon lease has been the core of the business and is expected to remain so for the next twenty years. Situated in the Dorsey area of the Mount Poso Field, the quiet foothills next to the mountains also provide a wonderful retreat, as well as a great oil property.

In 1987, while working with Stockdale Oil & Gas, Tony met Kurt Sickles and started a long and prosperous friendship. "Over the years, we owned and operated various properties through his Brea Oil Co., Inc.," Tony says. "The most memorable and important was the Sheep Springs lease in the Cymric Oilfield. In 2007, we drilled the Sheep Springs C-5 and that well flowed 100 BOPD for over a year. That was definitely one of the high points in both of our careers."

In summing up his career, Tony says, "The oil business has been an extremely important part of my life from day one. It has taught me many lessons and provided a wonderful life. I love the oilfields and am fortunate to have spent a lifetime in them."

MMI Services, Inc.

"Risk taker" is a term used often to describe Mel McGowan, and it is true he has known when to take advantage of opportunities and take calculated risks in order to move ahead. However, McGowan, who founded and still runs MMI Services, Inc., has been successful because he understands his business, values his customers, and has been willing to change with the times.

"I have always maintained three priorities for myself and our employees," Mel said in a magazine interview. "The priorities are safety, customer service and hard work. I have never found these three priorities to let me down."

MMI Services, headquartered in Bakersfield, is an independent well servicing contractor. The company provides quality service in a wide variety of completions, workovers, abandonments and production work for large oil companies, small independent producers and the State of California.

Mel grew up in the oil business and has spent most of his life working in the industry. He was born in Merced, California, the oldest of six children, but the family moved around a lot because his father did oilfield work, following the ups, downs and job opportunities wherever they took him.

Mel spent his first four years of elementary school in Taft, before the family moved on to Newhall. He left high school in his senior year in order to join the Air Force, though he earned his high school diploma later. Only a year after he joined the Air Force, Mel's father died of a heart attack at the age of forty-one. Mel was given a hardship discharge and came home to look after his family. He moved his mother and five brothers and sisters to Bakersfield, to be near his mother's relatives.

Not yet twenty years old, Mel began a series of oil-related jobs that ranged from working at a gas station to managing an oilfield supply business. The oilfield supply company transferred him to Ventura, but he did not like the change and returned to Bakersfield after only three months. Now unemployed, Mel went job hunting and found a position with a wireline company.

By 1970, Mel felt it was time to go out on his own. Although he had a family to support, including three young sons, he thought the timing was right. He sold his boat, borrowed $3,000 from a brother, and financed the purchase of a truck. The new business, incorporated as McGowan Services, had only one employee and a pick-up truck, but Mel was determined to build a successful company.

Mel's previous sales experience and on-the-job training was a tremendous asset and the company soon signed its first account with Chevron. The company expanded from there, adding more trucks and employees as the company grew. In 1975, Mel bought his first production rig and added additional equipment as the need arose. Eventually, the company owned four wireline trucks and around a dozen production rigs.

In 1985, McGowan Services, including all the equipment, was sold, although the new owners continued to operate on property they leased from Mel. Against the advice of his attorney, Mel financed the transaction, but did not do a UCC filing because he felt the buyers were strong enough financially that it would not be necessary. He did, however, retain the real property as an income investment. As it turned out, this proved to be a very wise decision.

Before long, Mel began to hear that the new owners were losing accounts he had built up and maintained over the years. The company was soon in trouble and the new owners asked Mel to lower the purchase payments. In an effort to save the company, Mel offered to return and manage the company. This offer was rejected, and Mel refused to lower the payments. Within a year, the new owners of McGowan Services were out of business.

"Most people think I just went in and repossessed the equipment and took over the business again," Mel says. "That's not true. I had to start over."

The new owners had borrowed heavily, using the equipment, rigs and trucks as collateral. The bank repossessed the equipment and shut down the operation, and Mel was left with only an unpaid note and no assets to back it up.

With the down payment from the sale of McGowan Services, Mel had begun buying and selling used oilfield equipment. When the buyers went out of business, Mel had two production rigs on hand he had intended to sell. With the revenue from selling those rigs, Mel was able to move back into his old office and attempt to revive the company he had founded.

A successful new company soon began to emerge from the collapse of McGowan Services. A new company, incorporated as MMI (Mel McGowan, Inc.), was formed and Mel began to rebuild the business he had founded with such bright expectations. Mel began to reestablish the business, adding wireline trucks and production rigs as it grew. By 1992 the company had 200 employees.

In the early 1990s, many smaller companies such as MMI found it hard to compete against the alliances formed between the big oil companies and service companies. Once again, Mel was willing to take a risk and move his company in another direction. MMI continued to provide wireline services but Mel decided to concentrate on well abandonments after the Division of Oil & Gas began to more strongly enforce the rules applying

RAIN FOR RENT

WESTERN OILFIELDS SUPPLY CO.

Above: Founder Charles P. Lake in front of Highway 99 Yard.

Below: Founder Charles P. Lake with a pipe-straightener. The company's pipe-straightening production line reconditioned bent scrap oilfield pipe into useable resale and rental products.

For more than eighty years, Western Oilfields Supply Co. (WOSCO) has provided clients with solutions to their liquid management challenges. This fourth generation, family-owned business has grown to become an International company with sixty-five U.S. locations and operations in Canada, Europe and the United Kingdom.

In 1934, Charles P. Lake and four partners founded the company in Bakersfield, near Hoover Camp, the shanty town made famous in John Steinbeck's novel the *Grapes of Wrath*. From 1934 until 1950, the company was almost exclusively in the oil and gas used equipment sales and rental business. Lake specialized in buying and reconditioning used oilfield pipe. "'Bought right, half sold,' was grandfather's favorite quote," says his grandson, John Lake.

Despite the economic depression of the 1930s, a number of independent wildcatters and oil companies were still operating in Kern County. They generated an ample supply of used pipe and equipment for the company to acquire.

WOSCO reconditioned damaged drill pipe by heat shrinking new sleeves onto old pipe and rethreading the couplings. Then, a spiral weld pipeline was added for manufacturing well casings. The company also bought, sold and rented temporary Victaulic, gas and water pipelines for the drilling industry in the San Joaquin Valley. Typically, the company located water sources for the drilling operators and provided temporary water pipelines for rig operations.

To survive during the Great Depression years, WOSCO built clothesline poles and purchased old steam locomotive boiler tubing to make 'shot tubes' for the explosives used for oilfield geotechnical seismic exploration. The company's pipe-straightening production line reconditioned bent scrap oilfield pipe into useable resale and rental products.

During World War II, WOSCO teamed with local manufacturers to build bulk-heads for Liberty ships. After WWII, local soldiers and sailors returned home determined to transform the San Joaquin Valley desert into the greatest farming region in the world. However, alkali in the soil was the enemy of farming and finding a way to reclaim the land became a significant challenge.

"Grandfather always said, never ignore reality, always use it to your advantage. Well, we had to do that after the end of the Korean War because the price of oil collapsed to less than $1 a barrel. Meanwhile, an extensive supply of aluminum from war time aircraft production became available for civilian use," explains John. "This aluminum could be used for a new product 'sprinkler pipe' and pipe was our business."

Fortuitously, Charles' son, Jerry, had returned to the business in 1948 after serving with the 8th Army Air Corps in England and he began developing the company's irrigation product lines, based upon aluminum pipe and sprinkler technology.

Because of the slow application of water via sprinkler irrigation, the alkali could be leached to below the root zones of the plants and the land could be reclaimed. The advent of pressurized irrigation systems sparked a revolution in farming and opened up millions of acres of former desert land all across the West.

When the company had difficulty selling the new sprinkler equipment to farmers, Charles tried renting it, an approach that had proven successful in the oil business. A new business name was needed for the agricultural products; and, with the aid of a local ad agency, the Rain for Rent trade name was launched. Millions of joints of sprinkler pipe were manufactured, sold and rented to farmers throughout the West, sparking the company's rapid growth into agriculture.

In 1957 the WOSCO foundry was established to cast irrigation valves and fittings. Later, in 1969, a tube mill was acquired to manufacture aluminum pipe for the rental fleet and resale. By 1976, new ag land development began to slow and two terrible droughts took their toll on the agricultural economy. "We began a search for new products and services to survive, using 'the reality of the situation' to our advantage."

Under the leadership of Don and Jerry Lake, the second generation, the company expanded the irrigation business throughout California, Arizona and Idaho, and supported Interstate highway construction projects across the West. Many new branches were opened and Lake Company, a wholesale division, was formed in 1981 to sell irrigation products to dealers nationally and internationally.

Upon Jerry's retirement in 1990 and Don's passing, the third generation took charge of the business. John, Cynthia and Robert Lake continued to diversify the company's products and services geographically. The company pursued opportunities in ten market segments: environmental, construction, oilfield exploration and pipelines, refineries, chemical plants, agriculture, mining, power plants, municipalities and manufacturing facilities.

The acquisition of Frac Tanks, Inc., in 1999 and the hydrofracking revolution changed the business again. Norm Stephen, former CEO of Frac Tanks, Inc., joined the company to lead the expansion of the frac tank business across the nation. New branches were opened in the Utica, Marcellus, Piceance, Bakken and other plays to provide equipment to major exploration companies. During the next two decades, water filtration, instrumentation, heat equipment, and spill containment product lines were added to the company's already extensive pipe and pumping capabilities.

Left: Jerry Lake started the irrigation business in 1948.

Below: The Lake family leaders. Three generations, left to right, Robert, John Paul, Cynthia, Jerry and John Lake.

Charles passed away in 1988 at the age of ninety-four and Jerry passed in 2013 at the age of ninety-one. They left a strong legacy of steward leadership and business values for the third and fourth generation Lake Family leaders to follow.

WILLIAM WARREN ORCUTT

DEAN OF PETROLEUM GEOLOGY

1869-1942

Written by
Mary Alice Orcutt-Henderson

Above: La Brea Tar Pits excavations in foreground, c. 1908.

Right: William Warren Orcutt and Mary Logan Orcutt.

Below: The Orcutt residence.

William Warren Orcutt was born in Minnesota. In 1882 the family moved to Santa Paula, Ventura County, California, where his father purchased a section of land east of the small town. Educated in local schools, Will matriculated to the new Leland Stanford Junior Farm in Palo Alto—the future Stanford University. By now a strapping young man of six feet and 190 pounds, he joined the football team where he earned a red "Block S" and in Track & Field he "put the shot" and "threw the hammer." Orcutt graduated in the Pioneer Class of 1895 with a degree in civil and hydraulic engineering.

Returning to Santa Paula, Orcutt married his childhood sweetheart, Molly (Mary) Logan, earned his surveyor's license and set up business in a small upstairs office in the Union Oil Company's headquarters. On occasion the young engineer was hired by the board of directors as a consultant. One report gives his telling analysis of their request to study the company's current operations. It is a typed, multi-paged evaluation, which both lauds and faults their practices and procedures.

Stunned by his astute appraisals yet impressed by his acumen of the industry and youthful enthusiasm, the men offered the thirty year old engineer a permanent position as general superintendent of their San Joaquin Valley division. The superintendent was directed to survey and supervise the construction of a crude oil pipeline from the company's fields to tidewater at Port San Luis in San Luis Obispo County.

It was during these weeks of surveying the route that "Orcutt the Geologist" emerged. He correctly assumed that by using visual observations and known geological presumptions that the finding of oil could be a more exact calculation rather than the typical hit or miss gamble of the day. Accordingly, he systematically applied his geologic theories to a hand-sketched

drawing of the great Santa Maria Basin while exploring the terrain by wagon, horseback, or on foot. The resulting map was the first charting of the formations, petroleum seeps, and unusual outcroppings indicating probable oil-bearing strata. It was the scientific approach he had been seeking and one that revolutionized the oil industry's exploration techniques.

He shared his map with his friend and colleague Will Stewart, who was so fascinated with the concept that he gave it to his father Lyman Stewart, who happened to be Union's current president. He and the board were so overwhelmed that immediately two motions were put up for vote: to buy or lease the approximately 70,000 acres delineated on Orcutt's map; wire him to quit Coalinga and come to their new headquarters in downtown Los Angeles. There was no hesitation on the young family's part.

The timing for Union was perfect. Relocating to Spring Street in downtown Los Angeles put the company in the center of the burgeoning business district and positioned it in the forefront of the emerging oil industry. W. W. Orcutt was authorized to establish and manage the new geological department—the nation's first. His probable well sites were proving out and new investors were pouring in.

To accommodate the ever increasing production of oil from new wells coming in and the number of Union employees and families moving in, a town site was surveyed and associated structures, infrastructures and facilities needed for a community was built according to the plan laid out by the engineer-geologist. In appreciation, his crew accorded the new town the name "Orcutt" in his honor.

Over the next thirty-one years Orcutt and his teams of eager young geologists explored unchartered lands in search of new oil fields. They hacked through jungles in South America and Mexico, stomped across the frozen tundra of Alaska, and rode horseback or drove throughout the western states. As a result the company acquired vast lands or leases, which proved to be among the richest in the world. Within the Los Angeles Basin their discoveries included Santa Fe Springs, Dominguez, Richfield and Montebello Fields.

In addition to his geological accomplishments, Orcutt is considered the discoverer of the La Brea Tar Pits because in 1906 he was not only the first to recognize that the bones were fossilized remains of animals from the Late Pleistocene Epoch but also, he initiated the scientific excavations. In the 1930s, Union acquired land that included the eroding adobe ruins of the La Purisima Mission near Lompoc. He was responsible for the company's deeding the original site to the Santa Barbara County for eventual restoration. In anticipation of Union's fiftieth anniversary celebration in Santa Paula in 1940, he organized the founding of the California Oil Museum on the first floor of the company's original headquarters. The museum continues to be a popular tourist attraction. Orcutt considered such activities all in a day's work.

The Orcutts had two children, Gertrude and John. Mary Alice Orcutt-Henderson, daughter of John shares, "When my grandfather retired in 1939, after thirty-seven years with the Union Oil Company, he had served on the executive board since 1908 and was the vice president of four divisions: geology, land, development and production. His decades of contributions to the industry lent to his recognition of 'Dean of Petroleum Geology.'"

Above: Orcutt Field, 1907.

Below: Orcutt hand-drawn map of Adams Canyon, Santa Paula, 1898.

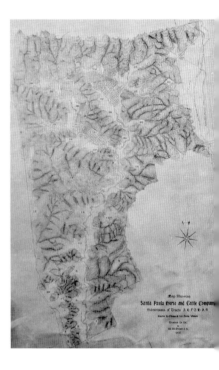

JONES GROUP

The Jones Group of companies, head-quartered in the Sacramento, California, suburb of Rancho Cordova, is a multifaceted energy company with deep roots in various sectors of the oil and gas industry since the early 1950s.

Above: The Jones family, left to right: Sandra, Derek, Griffin, Gloria, Vern, Erika and Alexys.

Below: The company's Gas Control Center.

Founded in 1952 by Stanford University geologist Vern Jones, Exploration Logging (EXLOG) provided well-site geological formation services, to the energy industry, known as "mud-logging." In the early to mid-1960s the company began international operations and by 1980, EXLOG was operating more than 200 field laboratories, or 'logging' units, employing more than 800 professional geologists, and providing services in twenty-six countries worldwide making it the world's largest mud-logging company. In 1972, EXLOG was merged into Baker Hughes where Vern remained at the EXLOG helm until his retirement in 1980, subsequently departing from the Baker Hughes Board of Directors in 1985.

Vern's retirement did not last long and in 1980 he and his wife, Gloria, began pursuing new activities with their son, Derek, and daughter, Sandra, including ventures in the wine business but more importantly, natural gas exploration. Those steps led to the formation of Vern Jones Oil & Gas, and the discovery and production of natural gas in the Sacramento Basin of California, as well as oil and gas exploration projects in Texas, Louisiana, Colorado, Kansas, and Oklahoma through the early 1990s.

The Jones' involvement in gas exploration in Northern California required new market access and pipeline infrastructure development that resulted in establishing many mutually beneficial business relationships with other producing companies and gas markets. Through these relationships, the Jones recognized an opportunity to extend their gathering, pipelining, and marketing of natural gas from wells being drilled by other producers. Long held to be the function of the local public utility, the combination of regulatory unbundling and the local utility exiting the gathering function, the door of opportunity opened for the Jones to form California Energy Exchange to develop new natural gas gathering pipeline and market connections capable of accommodating the nitrogen-rich natural gas often found in the region.

From 1985 to the present, the Jones Group has worked diligently to establish a strong presence in the energy industry both internationally and throughout Northern California, and more recently in Northern Nevada through Prospector Pipeline Company. Over the years these efforts led to the formation of Energy Operations Management (EOM) to manage the various companies and assets established as part of the Jones Group of companies.

In 1990 the Jones Group determined that new opportunities were emerging outside the United States and started International Logging, Inc. (ILI), to once again provide services to the oilfield industry. Started in Sacramento and Bakersfield in 1990, ILI grew to include over eighty logging units and nearly 400 geologists and engineers with headquarters in Houston, Texas and eight overseas offices with operations in Southeast Asia, China, Australia, Central and South American and West Africa. In 2006, ILI was sold to the Carlyle/Riverstone group and today is a part of Weatherford.

Recently, EOM has sharpened its focus to target natural gas projects in infrastructure-deficient areas where the economic and environmental benefits to its clients are the greatest and distinguishes itself by not only developing the projects but then going on to operate them on a long-term basis after construction. EOM's development of pipeline infrastructure today not only involves the technical challenges of the project but a thorough evaluation of the regulatory environment to establish the best business model for the project. EOM operations include the Gas Control Center, located in the Rancho Cordova headquarters, which monitors and controls the several hundred miles of natural gas pipelines on a 24/7 basis to ensure that the highest standards of operational safety and efficiency are maintained. EOM is fully compliant with the federal and state pipeline safety standards as administered by PHMSA (Pipeline and Hazardous Materials Safety Administration [federal DOT]) and by the PUCN (Public Utilities Commission of Nevada) respectively to ensure that standards for operations, maintenance, and record keeping are met.

The Jones Group's pipeline business has been headed by Derek, joined by his EOM Management Teammates including Chief Financial Officer Timothy Wise and Operations Vice President Brian Habersack. The team works with its highly-qualified staff and key contractors (twelve to seventeen individuals) to develop and manage its pipeline projects and to provide related services to natural gas end-user customers, primarily located in California and Nevada.

Experienced in pipeline development nuances of permitting, environmental, regulatory, legal, and operations, EOM can quickly assess the project priorities. Whether it is developing gas gathering pipeline systems, overcoming off-spec gas marketing issues, or providing complicated turnkey pipeline and natural gas supply arrangements, the Jones Group has the knowledge and experience needed to get the job done. Industry affiliations include the AAPG, IPAA, CIPA, and other industry organizations.

As the Oil and Gas industry in the U.S. continues to grow domestic resources, natural gas will play an ever increasing role as a 'cleaner' alternative in sectors such as manufacturing, electrical power generation, and for use as a transportation fuel. Armed with new strategies intended to meet the challenges of the energy demands of tomorrow, the Jones are facing the future with optimism—identifying new opportunities and forging ahead with a growth plan to meet the those opportunities that the future holds.

Above: The Prospector Pipeline Company twenty-four mile extension from the Ruby Pipeline in Nevada.

Below: Logging units being delivered to a remote Oxy location—the Arctic Circle, Russia.

AC PIPE & EQUIPMENT CO.

Alan Chavez

Above: Alan Chavez, president of AC Pipe & Equipment Co., and Terra Exploration & Production in 1993.

Below: Mick Conner, secretary/treasurer of AC Pipe & Equipment Co., and Terra Exploration & Production in 1993.

AC Pipe & Equipment Co. was organized out of necessity. The founder, Alan Chavez, was out of work and decided he had no choice but to go out on his own, selling used oilfield pipe and equipment left behind when oil wells were shut down.

Today, AC Pipe & Equipment Co. sells tubulars, structural casing and tubing, and all models and sizes of pumping units. The company has its own machine shop and provides a complete pumping unit service, which includes inspection, repairs, maintenance, refurbishment and installation. They also provide trucking services.

AC Pipe & Equipment Co. has come a long way in thirty years, from a one-man operation to more than thirty-five employees. It also has customers throughout the nation and in Korea, China, Russia and the Philippines.

The company was founded in 1986 after Chavez was laid off from a tubular company in Oklahoma. "Oil prices slumped in 1986 and since I was unemployed, I started selling all the surplus oilfield pipe and equipment I could get my hands on," Chavez says.

Chavez had little money to invest in his new venture and was forced to take out private loans with high rates of interest. But his contacts and hard work soon began to pay off and Chavez signed his first contract with Shell Oil Company. "In 1987, I was awarded a contract with the Exxon Wilmington field to remove over 400 pumping units and was involved with the abandonment of the wells," he says. "A year later, I received a contract with Chevron Santa Maria to remove 480 pumping units and tubing from their different leases."

Chavez also worked on the ARCO leases in Midway Sunset and Taft (now Linn Energy). "I was fortunate to have a successful relationship with THUMS of Long Beach, California, by acquiring pipe and other miscellaneous oilfield equipment and selling it to a company located in Texas," he explains. "This company purchased any and everything I had to sell and this became the catalyst for the growth of the business."

In 1989, Chavez teamed up with Michael Conner, who had also been laid off and was looking for a job. Conner was instrumental in selling new and used pumping units to all the oilfield companies in California, an effort that helped AC Pipe & Equipment Co. to grow and diversify. Conner was awarded forty-nine percent equity in the company in 1993.

"In 1993, Mick was out making sales calls trying to buy and sell pipes, pumping units and other oilfield equipment," Chavez recalls. "He came across a small oilfield in Culver City

that was full of weeds and non-operating pumping units. Mick contacted the owners to ask if we could buy the equipment and they replied that they were willing to sell the equipment as long as we purchased all of the wells that were on the oilfield."

This event resulted in formation of a second company, Terra Exploration & Production, which eventually grew to a total of 105 oil and gas producing wells.

Chavez admits that purchasing the oil wells with all their liabilities was scary. "Oil prices were low, and it was a gamble," he explains. "But I just had a gut feeling it would be good. Besides, I watched *Dallas* on TV and always kind of wanted to own an oil well."

Banks were reluctant to finance anything related to oil because the market at the time was so depressed. However, Chavez was able to borrow $50,000 at a high interest rate from a neighbor. After starting Terra, the company bought its fifth well for a forklift and numbers six through nine for $10. The tenth well was purchased for $5,000 and three more for $25,000. The high interest loan was repaid with profits from the pipe business. In a later transaction, Terra purchased a field in Colorado with fifty-four wells.

Today, AC Pipe & Equipment Co. employs thirty-five people and operates from a

seventeen acre yard in Bakersfield, which includes a 10,000 square foot machine shop.

AC Pipe & Equipment Co. and Terra are guided by strong values and take pride in their honesty and hard work ethic. There is also a high respect for employees and business partners. The companies are involved in a number of community activities, including Seal Beach Lions Club, the local API Chapter, Bakersfield College, Little League, AYSO, R. M. Pyles Boys Camp, and Petroleum Club of Long Beach.

The corporate office is located at 1250 East 23rd Street, Signal Hill, California, 90755. The office telephone number is 562-427-3633. The manufacturing and yard facility is located at 825 East White Lane, Bakersfield, California, 93307. The office telephone number is 661-836-9189.

Above: Left to right, Alan Chavez and Mick Conner with their workover rig in Signal Hill, California, in 1995.

Below: Left to right, Mick Conner and Alan Chavez with one of their oil wells in Signal Hill, California, in 1995.

PETROL TRANSPORT, INC.

For nearly four decades, Petrol Transport, Inc., has hauled crude oil for independent producers from the oil leases and rail cars into major pipeline systems in the San Joaquin and Santa Maria Valleys, which supply major California refineries on the central coast and in the San Francisco Bay area. Petrol Transport provides its customers with twenty-four hour service through its ability to maximize its fleet of thirty-two trucks at the most competitive rates in the industry.

Originally known as Flying B Trucking, Petrol Transport was created out of a friendship between Ted 'Brownie' Barnard, Jr., and his college classmate Dennis Burtch. In 1979, Burtch took over his father's company, Burtch Trucking. The company was hauling oil at a lease in the Poso Creek area, however, Burtch was concerned that his crews were falling behind and that his company would lose the haul as a result. Burtch reached out to Barnard, who was working as a ranch foreman for Joughin Ranch, after graduating from Cal Poly,

and asked if he would help him out by observing the lease for a week, to determine the problem and find a solution. After a few days, Barnard felt that since a new crew was sent in for each job, each new crew was repeating the mistakes made by the previous one and suggested that Burtch put a dedicated crew on the haul, figuring that they would make mistakes and learn from them.

At the time Burtch could not dedicate a crew to the lease and asked Brownie if he would like to purchase his own truck and operate under Burtch Trucking, Brownie agreed, forming Flying B Trucking from that one truck.

In the 1970s drivers were not required to attend a truck driving school to obtain their Class A driver's license—a designated company official could sign off on a new driver so long as he felt the driver was proficient. Brownie's first haul consisted of hauling one load of fuel oil from Bakersfield to Long Beach, a trip filled with challenges and excitement. He was driving an older, 'needle nose' Peterbilt truck, which had two sticks—a main gear box with five gears and an auxiliary gear box with an additional four gears. The haul was made even more difficult because Brownie had to drive over the 'Grapevine,' a treacherous 200 mile stretch of Interstate 5 from Los Angeles to Kern County, during a snow storm. Upon arriving at his destination, the fuel oil had become cold and viscous, taking twelve hours to offload. By the time he was back on the road, the Grapevine was closed due to the storm and he had return via the coastal route to get back to Bakersfield. With all the delays, that initial load took twenty-four hours to complete. Once he arrived back in the yard, the vice president of Burtch

Trucking signed off on his training, commenting that in a twenty-four hour period he had experienced more than he would have in a year of training.

For the next three years, Brownie did double duty, driving his Flying B truck at night, and working as a ranch foreman during the day. This grueling schedule meant plenty of twenty hour days. In 1982, Brownie learned that Burtch was interested in selling his trucking company, and two of Burtch's vice presidents as well as Brownie were interested in purchasing it. A short time later, during a fishing trip to Canada, Brownie saw his opportunity to make a pitch to Burtch. Brownie gave the fishing guide a generous gratuity to make sure he and Burtch were on the same boat. This resulted in the 'handshake' deal that allowed Brownie to acquire the company for a down payment and monthly payments over the next five years. The purchase increased Flying B's fleet size to sixteen trucks and the company name was then changed to Petrol Transport.

Under Brownie, Petrol Transport grew and was very successful—so much so that in 2009 another trucking company expressed interest in buying the firm. When Brownie's son Zach Barnard, who was working for an inspection company in New Mexico, after his discharge from the Navy, heard of the possible sale, he returned home. "I couldn't imagine Petrol Transport being sold and so I made the decision to come home and work in the trucking business," said Zach. The younger Barnard eventually became a partner in the company in 2010. Since that time, Petrol Transport has continued to grow and increase its customer base, and is now one of the largest and most highly regarded crude trucking companies in the Southern San Joaquin and Santa Maria Valley regions.

The company's fleet has grown to 32 trucks and employs 80 people, 65 of them are drivers. Long-time employees who have been influential in the company's growth include Clyde Blackwell, who worked as a truck driver, dispatcher, and then general manager, until 2007, and Roy Crawford, a dispatcher who was with Petrol Transport until 2005. The company has main offices in Bakersfield with satellite offices in Santa Maria and Coalinga, and plans for additional locations in the near future.

Among the company's key customers are California Resources Corp., E&B Natural Resources, Kern Oil & Refining, LINN Energy, Plains Marketing, Phillips 66, and Vaquero Energy.

Zach continues to operate Petrol Transport by the principles under which his father started the company four decades ago. Zach explains that principle is summed up in something Brownie always said to him, which was, "'I never wanted to be the biggest, I only wanted to be the best.' That philosophy has guided the growth of Petrol Transport for thirty-six years," says Zach. "We have lived up to that principle, just ask our customers, and we will continue to do so for the next forty years."

JD RUSH COMPANY, INC.

Right: Burl Varner.

Below: J. D. Rush.

For an oilfield service company to complete eighty-two years, withstanding mergers, acquisitions or attrition, is an accomplishment of note. However, it becomes even more significant when it encompasses eight decades of family management. JD Rush Company, Inc., is now headquartered in the heart of Central California and holds a commanding position in the Oil Country Tubular Goods (OCTG) market. JD Rush Company has divisions throughout the United States and maintains an OCTG service center that spans more than 100 acres and includes a family of five related businesses.

The company was founded in 1933 when J. D. left General Petroleum Corp., where he had been a warehouse manager, and set up JD Rush Company in the Vernon industrial area of Los Angeles. His son-in-law, Burl Varner, soon joined the company that was selling general oilfield equipment, new and used pipe and related piping products, primarily in the Los Angeles Basin area. The business prospered and, in 1946, a move was made to a seven acre site in Gardena, California.

When J. D. retired, Burl became a company principal and eventually was named company president. Jim Varner, son of Burl and grandson of J. D., representing the family's third generation, joined the firm in 1966 and rose to become president and CEO in 1974. It was during this time that JD Rush Company began to experience tremendous growth and rose to

prominence as a major force in OCTG distribution. JD Rush Company reached a distribution agreement with NKK to stock and sell OCTG in 1975 and became a US Steel Distributor in 1985. These world-class mill alliances and savvy management helped solidify JD Rush Company's position as a regional powerhouse. Today, Jim remains CEO while his daughter Teri Seely, representing the family's fourth generation, is president. Teri's son and also the great-great grandson of J. D., Taylor Dickson joined the company full time in 2014.

JD Rush Company has a history of capitalizing on downturns and this was certainly the case in the 1980s. The company set out on a task to differentiate itself from their competitors by adding value through a "one stop shop" facility that has received continual reinvestment, additions and improvements through the years.

As the oil and gas industry expanded and contracted over the years, JD Rush Company developed new products and services to meet customer's changing needs. By the mid 1980s, it became clear to company executives that the best way to serve its customers was to expand its capabilities. The goal was to cut costs and add value, and the best way to do that was to build an integrated service center where value added services could be conducted at one location. Today, JD Rush Company represents both domestic and foreign seamless and ERW pipe manufacturers.

The family of domestic JD Rush Companies includes Tryad Service Corporation, West Coast Pipe Inspection, Trymax International, and JD Rush Corporation. These companies provide services to independent and major oil companies throughout the North American market.

Tryad Services Corporation, established in 1953, is an American Petroleum (API) Certified threading shop based out of Shafter, California. Tryad is a world-class supplier of down-hole requirements for the international oil and gas industry. Tryad was granted a U.S. Patent for its One Trip Teflon seal Adapter® in 2004.

Founded in 1987, West Coast Pipe Inspection and Maintenance, Inc., (WCPI) is headquartered in Shafter where it maintains more than seventy acres of pipe storage, an onsite rail spur, and is easily accessible to California's highway systems. With a fleet of mobile inspection units, WCPI can easily accommodate each customer's schedule and needs.

Trymax International, Inc. was created in 2001 as a joint venture between the JD Rush Company, Inc., and MC Tubular Products, Inc., (a subsidiary of Metal One Corporation). Originally, the company's primary purpose was to produce the SuperMax thread, commonly known as SMAX. Trymax also offers American Petroleum (API) Certified threading up to seven inch OD, digital torque turn, propriety threads including the GeoConn connection as well as premium options.

Established in 2003, JD Rush Corporation is a joint venture of JD Rush Company, Inc., and MC Tubular Products. Headquartered in Houston, Texas, JD Rush Corporation has sales offices in Dallas, Denver, and New Orleans and maintains more than thirty strategically placed stocking locations across the United States. JD Rush Corporation's experienced personnel are ready to meet the service requirements of the oil and gas industry from procurement, threading, and inspection to logistics.

By contracting with affiliated Weststar Trucking, JD Rush Company is able to supply customers with personalized transportation services at reasonable rates.

At JD Rush Company, the entire family of affiliated and associated companies is committed to maintaining an environment of continual improvement and safety through its quality control and safety plans. The company emphasizes defect and variation prevention as it strives to reduce waste and maintains an extensive safety program with employees regularly participating in training and development programs.

JD Rush Company believes in giving back to the community and participates in numerous charitable and community programs.

For information about JD Rush Company, check the website at www.jdrush.com.

Left: Jim Varner.

MUTH PUMP LLC

*Above: Left to right, Garold and son
David Muth next to the demonstration unit
in the trailer.*

*Below: Garold (far left) and son
David Muth (right of Garold in black shirt)
demonstrating the demo unit to Aera
Energy engineers.*

It might sound like a Sunday school story from the Old Testament, but Garold Muth, inventor of the FARR Plunger, has no doubt that the idea for his pump was nothing less than a revelation from the Lord. After hearing the story, you will probably agree.

The FARR Plunger is a rod pump plunger that has revolutionized the oil industry and how production companies deal with sand and other solids. Although the inventor believes the product was divinely inspired, it took years of hard work and determination before it was fully accepted in the industry.

Like many in the industry, Muth grew up in the oil business, following in his father's footsteps. After growing up in Ventura County and serving five years in the U.S. Navy,

Muth returned home in 1963 in need of a job. He learned that Texaco was looking for a roustabout in Los Angeles and he got the job, beginning a long career in oil production.

It was 1990 when Muth had the dream that changed his life and led to creation of the FARR Plunger. As he explains it, his wife, Mandy Muth—a real estate broker—kept a notepad and pencil on the bedside table so she could write down information from clients who called late at night.

"I had a dream one night, but had forgotten all about it," Muth explains. "While I was eating breakfast, my wife came in with a sheet from the bedside notepad that was filled with diagrams and hard-to-decipher notes. It was then that the dream came back to me in full detail.

"In the dream, I was a young man in the oil fields, driving along in my pick-up truck and checking wells. But, every time I drove up to a well, the pump would be down. I thought to myself that there had to be a better way and all of a sudden the light bulb came on. Apparently I was conscious enough to jot down some drawings and notes on the bedside pad."

When Muth looked over his notes, the pump he had dreamed of in his sleep actually made good sense and he was determined to develop the product. With conventional plungers, when sand enters the production string, the sand will get between the plunger and pump barrel wall and stick the pump; stopping production. Muth was convinced his divinely inspired product could solve the problem.

Muth persuaded his wife to support the family with her real estate business while he devoted full-time to developing the pump that had been revealed to him in a dream. He set up a shop in his garage, refined his drawings and obtained a patent on the product. It was named the FARR Plunger in honor of his father-in-law, Jimmy Farr, and also because the name lent itself to a great advertising slogan, "By FARR, We Make Your Rod Pumps the Best in the Industry."

By 1995, Muth was able to formally organize Muth Pump LLC and moved into a small 1,000 square foot warehouse on New Horizon Boulevard in Bakersfield. Production of the product was outsourced and Muth devoted his time to promoting his revolutionary pump.

Although the pump lasts 300 percent longer than others and lowers operating expenses for the oil producers, it was tough for Muth to break into the market. "I had to fight the big guys," Muth explains. "When the big manufacturers found out about the product, they put a hex on us. They all had standing relationships with all the big oil companies that went back years and years, so it took about ten years before the product caught on."

It was not until a friend, Roger Ross-Smith, built a working model of the plunger that the oil companies gradually warmed to the idea. "The big breakthrough came when an engineer with a prominent Kern County oil company, took a liking to it and pushed his company to buy a few of the pumps. In order to dismiss the hex the big manufacturers had placed on us, the prominent oil company conducted a one-year Six Sigma Study of the FARR Plunger pump in comparison to the big boy's pumps. The study found that the FARR Plunger pumps increased run times 300 percent over the competitors. It all took off from there," Muth says.

The FARR Plungers unique patented design allows it to pump when sand and other solids are present in the production fluid, without sticking. It can be used in any API pump barrel for insert and tubing pumps.

On average, the FARR Plunger will pump three to six times longer than conventional plungers. In addition, the FARR Plunger reduces environmental spill incidents, saves

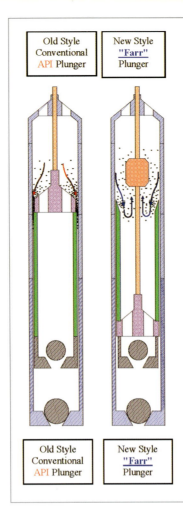

Old Style Conventional API Plunger	New Style "Farr" Plunger

Old Style API Plunger

The connecter at the top of the old style plunger that connects the plunger to the pull rod has a 60 thousands gap between itself and the pump barrel wall. It is this connecter, with its gap, that forces the coal fines and sand down and outward between the plunger and pump barrel on the up stroke. This is what will stick your pumps.

New style "Farr" Plunger.

The connector, which creates the problems in the conventional pump, has been moved from the top to the bottom of the plunger in the "Farr" design. This simple change eliminates the gap at the top and moves it to the bottom where it does not matter. Also, the angle at the top of the "Farr" plunger has been reversed to force sand inward as opposed to outward. This forces the coal fines and sand inward where it will catch the fluid flow coming up through the inside of the plunger and putting it back into solution. Now, the top of the "Farr" plunger has only a 2 or 3 thousands clearance between the plunger and pump barrel as opposed to the 60 thousands in the conventional design. We have closed up the gap by ninety seven percent (97%). That means 97% less chance of sticking the plunger and longer run life.

energy because of less frictional drag, reduces health and safety incidents, and reduces wear on rods and tubing. All of this results in increased production and greater profits.

Muth Pump now operates from a 9,000 square foot facility on Alken Street in Bakersfield where the company provides complete pumps with FARR Plungers locally and globally.

"Without God, none of this would have been possible," Muth firmly believes. "I attribute all my success to the Lord and the dream he gave me twenty-five years ago."

Visit the website at www.muthpump.com for detailed information on studies of the FARR Plunger.

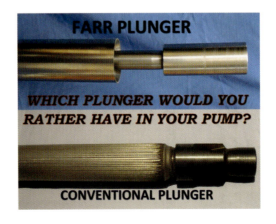

FARR PLUNGER

WHICH PLUNGER WOULD YOU RATHER HAVE IN YOUR PUMP?

CONVENTIONAL PLUNGER

KENAI DRILLING LIMITED

Kenai Drilling Limited, the largest independent drilling contractor in California, provides high quality drilling rigs with exceptional performance. Customers choose Kenai because of its many long-term relationships and its commitment to safety, quality and service.

Kenai was formed in 1988, initially with four drilling rigs, to specialize in the California intermediate and shallow market. Kenai has grown and expanded its rig count over the years by purchasing and building new rigs in order to meet customers' needs for the deeper market. Kenai now has 20 drilling rigs—12 in California and 8 in the Mid-Continent region—capable of providing drilling services for exploration and development of oil, gas, and geothermal wells. Kenai has the technical manpower necessary to provide any services a customer may request, from planning and drilling of wells, to building site specific equipment for hard to drill areas. Kenai is also capable of providing manpower for offshore or onshore projects while operating a customer's equipment.

Kenai was organized nearly twenty-eight years ago when an investor group headed by native Californian Tim Crist purchased Kenai Drilling from Kenai Corporation. The company was renamed Kenai Drilling Limited and over the next four years, Tim purchased the outstanding shares of the company from the original investors and became sole owner

Above: Kenai Rig #2 on Venoco Platform Gail.

Below: Kenai Rig #6 working near Blackwells Corner, California.

of the company. Kenai continues to remain as a privately held family business enterprise.

By the time he received his B.S. degree in Business Administration from California State University at Fresno, Tim had already accumulated five years of contract drilling experience. He began working full time for Kenai Corporation immediately following his graduation in 1980. Tim rose through the ranks and was named president of Kenai Drilling in 1985. He then led the successful effort to purchase the drilling assets from Kenai Corporation.

Other key individuals who have been deeply involved in the development of the company include David Arias, Carl Hathaway and Rex Northern.

David started his career in the oil and gas industry with Camay Drilling in 1977. In 1981 he transitioned to the oil and gas industry service sector. David joined Kenai Drilling in 2011 as the business development manager focused on industry relations and business growth. In May 2012, David was promoted to the position of executive vice president/COO with responsibility focused on California Operations and Business Development. A combination of David's thirty-five years of California Oil Industry experience and the relationships he has built throughout the industry have been proven to be a great addition to Kenai's already strong team.

Carl's forty-five years of experience in all aspects of the oil industry have been a great asset to Kenai. Carl graduated from Cal Poly San Luis Obispo in 1974 with a degree in mechanical engineering. He worked for Getty Oil Company in various engineering capacities for ten years. After working in engineering management for four years, Carl joined Texaco and worked several years in its environmental department. During that time, he worked on Platform Harvest in the area of regulatory compliance. Carl became Kenai's Safety Coordinator in 1995.

Rex started in the oilfield industry as a human resources and safety manager for an oilfield trucking company in 1990. In 1992 he accepted a safety manager position with JSM Drilling in West Texas, and then worked

for Nabors Drilling USA in West Texas and Bakersfield, California from 1997 to 2004 in the safety and HR departments. In 2004, he worked for WESTEC (Taft) as program manager at Drilling & Well Service Training Schools. In 2005-2006 Rex provided safety consulting in various areas of Safety/Training/Fall Protection/BBS/Equipment Integrity. In November 2006, he was offered an HES Specialist-Drilling position with Oxy Elk Hills, and was a member of the Oxy Global Drilling Safety Team until February 2008. In March 2008, Rex accepted a position as vice president and general manager of Safety & HR with Kenai. As part of the Kenai Management Team, Rex is charged with roles of responsibility in safety and human resources and is working on some new programs that will be beneficial to all in the coming years.

Kenai's corporate offices are located in Santa Maria. The company also has operations offices in Bakersfield and Liberal, Kansas.

Kenai is an active supporter of a number of charitable organizations throughout California and the Mid-Continent. These include American Cancer Society; Ventura and Kern County Youth Livestock Auctions and annual fundraisers for 4H and FFA programs; and Pyles Boys Camps. Kenai has also been one of the California Independent Petroleum Association's (CIPA) top sponsors for several years.

Few companies can equal Kenai's record in hiring veterans. The company believes strongly that veterans and servicemen and women are owed a debt as a result of their service and that they make great employees. The company operates a free program that connects job seeking veterans and military personnel with other employers that, like Kenai, have agreed to hire military members. Kenai always puts qualified veterans at the top of the list when reviewing applications and has been able to fill most of its vacancies with vets and members of military reserve units. In 2014, Kenai hired more than forty employees with military experience.

Kenai's high standards and job excellence earned the company the Drilling Company of the Year Award in 2013 from the California Independent Petroleum Association. The company has also been recognized by the International Association of Drilling Contractors (IADC) for its companywide safety programs and results.

Kenai Drilling Limited has the technical manpower and equipment to provide any service a customer may request. With this capability, Kenai intends to continue to provide safe and efficient drilling services and to become the driller of choice for all operators in the California market. The future looks bright for Kenai and the company feels it is well positioned to take advantage of the market when oil prices rebound.

Above: Kenai Rig #7 working near Bakersfield, California.

Below: Kenai Rig #10 at sunset.

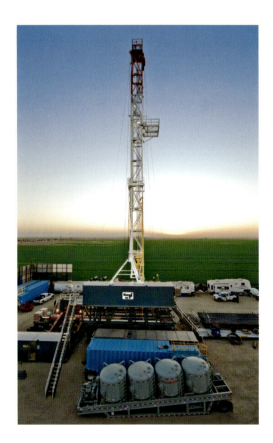

INTERACT PMTI, INC.

InterAct PMTI, Inc. is an international project management, engineering and environmental consulting firm specializing in oil and gas production, development, and decommissioning services.

Headquartered in Ventura, California, InterAct is made up of experienced, solution-minded engineers, scientists, environmental, and safety professionals covering a wide range of disciplines and expertise in the energy market.

InterAct's personnel are safety oriented, dedicated and highly skilled. The firm's professionalism and passion for helping clients achieve their goals, along with its interdisciplinary expertise, has enabled InterAct to become a leader in its field.

Founded in 1995 as Fairweather Pacific, LLC, the company successfully managed oil and projects in one of the most highly regulated environments in the world. The company later changed its name to Pacific Management Technologies, Inc. (PMTI), continuing its mission of providing engineering, environmental, and field support for oil production and end-of-life decommissioning projects in the Pacific region and Gulf of Mexico.

The Acteon Group acquired the company in 2007, changing its name to InterAct, reflecting a key aspect of the company's winning formula—effective interactions with clients, regulators, vendors, and subcontractors for successful project management. Acteon continued to expand its portfolio with a focus on engineered solutions between the interface of offshore wells and production facilities throughout the oil field life cycle. Over the last decade, the Acteon Group has grown nearly ten-fold into a global subsea services company.

Being part of Acteon enables InterAct to draw on a large international pool of skills and expertise to complement those found in-house. Working together, InterAct has completed projects in oil and gas markets around the world including well conductor removals in the South China Sea, well abandonment programs in the Middle East, well casing design in the Mediterranean, and decommissioning planning and environmental work in West Africa and the North Sea. This is in addition to servicing its home market clients in California, Alaska and the Gulf of Mexico with their engineering, operations, and regulatory challenges.

The motto at InterAct is: We plan. We engineer. We deliver. *You Succeed.* InterAct's experienced personnel offer practical, well-conceived solutions to client-specific needs, knowing that safety and environmental stewardship are cornerstones to success.

InterAct assists clients with all phases of oil development from acquisition due diligence and feasibility assessments, to drilling, facilities construction, and field and reservoir optimization, and final facility abandonment. The range of expertise makes InterAct an ideal, ready-made contract operations partner for oil field investors.

InterAct's offering includes:
- Activity Management—InterAct project managers include experienced engineers and operations personnel with a keen understanding of the metrics and drivers for on-time, on-budget project execution.

- Engineering and Field Services—InterAct provides comprehensive drilling, production, reservoir and facilities engineering, along with operational field development and management services. Additionally, the company employs drilling and workover site managers as well as API-trained crane operators and riggers.
- Environmental and Regulatory Services—InterAct offers comprehensive environmental and regulatory services including permit acquisition, environmental impact assessments, and regulatory compliance support.
- Decommissioning Services—A recognized leader internationally in this field, InterAct provides cost estimation, permitting, planning and execution services for facility removals and well plugging and abandonments.

The firm helps clients achieve their project goals in strictly regulated environments. With new interpretations of environmental laws and the promulgation of new regulations, InterAct helps companies understand and develop practical programs for implementation. Examples include land use permitting, groundwater protection through zonal isolation analysis, air quality and greenhouse gas emissions calculations, quantitative risk assessments, process safety analyses, industrial stormwater permitting, and spill prevention and contingency planning.

InterAct's technical staff have performed well integrity reviews for more than 8,000 wells and developed cost models and procedures for well plugging and abandonment and facility decommissioning. Its project managers have managed offshore rig moves, crane replacements, platform retrofits, and major onshore plant construction projects.

InterAct works across the industry, successfully facilitating multi-operator challenges and finding coordinated solutions. An example of this was the Subsea Well and Rig Sharing (SWARS) program—a six-company campaign to mobilize a jack-up rig originating from West Africa to abandon twenty-three subsea wells in the Santa Barbara Channel. The program was a success by all measures, as it efficiently and cost-effectively accomplished the intended goal of permanently plugging the wells while reducing environmental impacts. The company is embarking on a similar approach with its Pacific Abandonment Coordinated Enterprise (PACE) program, and is assisting several companies with the planning of the eventual decommissioning of west coast offshore platforms.

InterAct and its employees are committed to giving back to the community and are involved in a number of civic and charitable activities, including college scholarships through the API, fundraisers for cancer research and the Boys and Girls Clubs of America, as well as helping to fund energy education initiatives.

InterAct and its professionals are active members of numerous professional engineering, environmental, and management associations. The company is also registered with PEC and ISNetwork. Find out more at www.interactprojects.com.

RAMSGATE ENGINEERING INC.

Ramsgate Engineering, Inc., of Bakersfield is a group of engineers, geologists, project managers, designers and support staff who design, procure, manage and operate all phases of thermal and related oilfield projects.

Ramsgate was founded in 2006 by five partners who had previously worked together at Getty Oil Company in the 1970s. Getty was subsequently purchased by Texaco, and Texaco was then purchased by Chevron. The five founders went in different directions through the years, but ended up back together to start Ramsgate.

The owners of Ramsgate are President Don Nelson, Engineering Manager Eric Berger, Mechanical Engineer Frank Lawrence, Controls Engineer Mike Houghton, and Construction Manager Danny Henderson.

The company's first project was development of a new, large-scale steam pilot project in Kuwait. Completion of the project took about two years and the firm continues to support the project.

Several of the engineers at Ramsgate are pioneers of the heavy oil industry and have developed many of the processes and equipment required to produce heavy oil.

Ramsgate provides a wide range of heavy oil development and enhanced thermal technologies, including steam generation systems service, repair upgrades and organization; water treatment for boiler feed and other purposes; crude oil, water and gas measurement and processing; and waste fluids management and disposal systems. Other services include the design, specification and optimization of heat exchange systems analysis, optimization, design and construction management of electric power systems; equipment and material procurement and many others.

Ramsgate is also affiliated with ProGauge, (sister company) a packaged equipment manufacturer dedicated to custom thermal and related processing packages. ProGauge custom equipment can generally be manufactured within sixteen to twenty-four weeks of order, when standard materials of construction are specified.

Among the products offered by ProGauge are steam generators, well testers, well selection manifolds, steam manifolds and fluids conditioning heat exchangers.

Ramsgate operates from three office locations in Bakersfield and has registered foreign branches in the countries of Oman and Bahrain.

The company has approximately 100 employees and has exported more than $125,000 in equipment around the world. Ramsgate has grown about fifteen percent each year and plans are in place to continue this growth pattern.

The Export-Import Bank of the United States (Ex-Im Bank) has appointed Don Nelson, president of Ramsgate Engineering and ProGauge Technologies, to the Export-Import Bank of the United States Advisory Committee. In its thirty-second year, the Congressionally-established Advisory Committee advises Ex-Im Bank on its policies and programs, in particular on the extent to which the Bank provides competitive financing to support American jobs through exports.

Bottom, left: ProGauge (Ramsgate) steam generators in Oman.

Bottom, right: ProGauge Size 1 Well Test System.

The heavy oil industry in Bakersfield has been around for quite some time, but other nations, specifically those in the Middle East and South America, are just beginning to embark on heavy oil development. In fact, Nelson says that Ramsgate has often been the lead engineer on a country's first heavy oil development.

"In Bahrain, for instance, we developed their first steam project in 2010," Nelson says. "We also did three different pilot plant projects at three different fields in Oman. They were so happy with the pilot projects that we did a larger plant for them and also manufactured their steam generators and well testers. They also had ten engineers come from Oman for a year-and-a-half to learn from us."

Other Middle Eastern nations Ramsgate has worked with include Kuwait, Saudi Arabia and Egypt. It also has worked in Indonesia, Madagascar and Albania. Today, the company is seeing South America's heavy oil development emerge.

"We are working on a project in Brazil to build steam generators, and Colombia has requested engineers to come there once a month to help define a project, so it looks like we're going to be doing large projects in South America in the near future," Nelson says. "It looks like the heavy oil industry is starting to ramp up there and I expect that we will be in the middle of it."

Each country Ramsgate enters will benefit from the company's extensive knowledge of best practices and industry trends. For instance, standards for emission controls have become increasingly stringent and Ramsgate has developed technology that meets and often surpasses industry guidelines.

"We've really optimized the controls on our equipment, such as steam generators, to control the emissions," Nelson says. "In the '70s and '80s, the emissions coming out of steam generators were really bad; they put out a lot of pollution. Now, the industry has them very optimized and very tightly controlled."

Ramsgate takes these insights and uses them in the unique designs requested by its clients. Not only is the company a one-stop shop, but it is also a build-to-order operation that can design, build and operate a project from scratch.

"We are a true solutions provider, not just a job shop," Nelson says. "We are very specific to what types of projects we do and are very good at helping our clients get the solution that's correct for them."

Ramsgate/ProGauge and its employees have supported many charitable and civic organizations through the years, including Frontier High Titan football, IEE China Lake Section, Frontier High Marching Band, Centennial High School, various school choirs, Kern County Fair Junior Livestock FAA, Bakersfield Homeless Center, American Legion, Boys and Girls Clubs of Kern County, Kern Economic Development, Kiwanis Club of Rosedale, North High Track, and many others.

Ramsgate Engineering's business plan for the future is very simple—continue to provide the best in class engineering services and equipment, and focus on Enhanced Oil Recovery projects around the world.

Bottom, left: ProGauge Size 3 Unit in South America.

Bottom, right: ProGauge Well Selection Manifold Module with flowback measurement and control option.

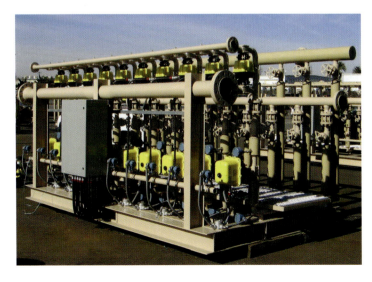

PETROLEUM PRODUCTION TESTING, INC.

Petroleum Production Testing, Inc. (PPT), the oldest production testing company in California, has served the oil industry for nearly thirty-five years. PPT offers a full line of production equipment, as well as some of the most experienced people in the test business.

PPT tests oil and gas wells, including acid flowbacks, frac flowbacks, free flowing wells, and wells that require secondary lift systems. For exploration wells, PPT provides test labs and personnel to monitor the wells twenty-four hours a day, seven days a week. The firm also provides daily production testing services that help customers calculate the daily production of oil, water and gas rates.

PPT was founded by Michael Poelke and his wife, Mary after Michael received some valuable but unexpected on-the-job experience. Although, he had been working as a real estate broker he felt it was time to find a different career. Since he was interested in the oil business, he talked with the owner of a well test company but was told there were no openings at the time. However, a current employee received an unexpected DUI and two days later Michael received a phone call asking him if he wanted to go to work in the oil business.

"Coming out of a three-piece suit and entering the oil business was going to be quite a change," Michael recalls. "This was the beginning of my oil field career but little did I know where it would take me."

Despite his lack of experience, Michael was put to work on the repair rack. "But, after about two weeks, my boss told me to go home and pack my bags because he was going to take me up north to start testing wells.

"Turns out the job site was the largest onshore find in years," Michael continues. "When we arrived it was dark and we were in the middle of nowhere. All I saw was a big flame coming from the end of a pipe, making an eerie glow in the night sky. My boss just looked at me and said, 'It's time to sink or swim.' Then the boss said he'd see me in three weeks and drove off."

The only other employee on site left shortly after Michael's arrival, leaving him alone and bewildered. "That was when the real panic set in," Michael recalls. "There was about 2,000 psi on the wellhead and the well was flowing into a two-phase separator. I stood there thinking what had I gotten myself into now." Then I headed for the test lab and saw a log with a lot of numbers on it, but I didn't have a clue to what those numbers meant."

Michael was left to figure things out for himself and was on the job site twenty-four hours a day, seven days a week for eight months, managing to go home only half a dozen times. He was a quick study, however, and his intense on-the-job immersion in the business was a great learning experience.

Oddly enough, Michael started his own business because he was fired from the job after he refused to charge a good customer to move a test lab that was only a quarter of a mile from a new well test.

Michael had made some good contacts at several oil companies during his short time in the business. The vice president of one of the companies promised to let him do his testing if he would start his own company. He and Mary founded Petroleum Production Testing, Inc. in 1981. "She handled the financial side of the operation, I handled the field operations," Michael explains. "In the early days, I tested most of the wells by myself and was on call twenty-four hours a day. Occasionally, when we could afford it, I called a temp agency that would send someone to help. I usually told them to watch a pressure gage and wake me if it moved. That way I could sneak in three or four hours of sleep."

PPT's growth was slow and methodical but the new company thrived despite the volatile nature of the oil business. Michael and Mary's three sons, James, Eric and Joshua, all worked for the business over the years and the youngest son, Joshua, is now a vice president of the company. PPT employs from ten and twenty-eight employees, depending on the current business environment. "We have chosen to stay a small- to medium-sized company," Michael explains. "This way, as owner, I am better able to offer our customers quality service."

PPT, which is located in Bakersfield, has been involved in a number of community activities. When gas prices reached record highs a few years ago, Michael realized that some people could not afford to keep gas in their cars. He contacted a local radio station, persuaded other oilfield service companies to participate, and sponsored a $100 gas card giveaway to the public every hour for a two week period. The company also supports the American Cancer Society and various school fundraisers.

EMJAYCO, L.P.

Emjayco, L.P., which evolved from Trico Oil & Gas, got its start when Harry Magee went broke during the Great Depression.

A third generation Californian whose ancestors had come from Ireland, Harry gained a reputation as an unconventional risk taker at an early age. He was kicked out of Berkeley and never finished college but filled his time as a daredevil motorcycle and speed boat racer.

As his grandson, Jed Magee, explains the story, Harry had become a very successful commercial real estate broker in San Francisco until he lost the business in the aftermath of the 1929 stock market crash that precipitated the Depression.

"Granddad was kind of a wild man and still had a lot of contacts with old-money people in San Francisco, so he approached them and said, 'Well, let's go drill some wells.' So, he traveled down through the San Joaquin Valley, looking for possible sites to drill, not really knowing if there was anything there or not. Granddad was a big hunter and always had his dogs with him, and family lore has it that he staked a well where one of the dogs stopped to take a leak."

The well turned out to be the discovery well for the Trico Gas Field, which allowed

Harry to get back on his feet and establish himself in the oil and gas business. With the backing of financier Elmer Stone, Superintendent John Qualman, drilling and production men Slim LeGar and Frank Lamb, geologist Wes Porter and others, Trico Oil & Gas was organized in 1932.

The oil business boomed during the war years of the 1940s and in the 1950s, Harry continued to demonstrate his magic touch for finding new fields. Defying conventional wisdom that there was no oil east of the main Mount Poso Fault, Harry decided to drill anyway. The first attempt was a dry hole, but the second established production east of the fault with the Glide 15 lease, which is still producing today.

Not all of Harry's ventures, however, turned out to be a winner. In a wildcat venture near Hollister in the early 1950s, Harry drilled a well that came in around 1,000 bbls per day. "Within weeks, a guy from Texaco showed up, wanting to buy the lease," Jed recalls. "He placed a blank check on the table, telling Harry to 'just fill in the number.' Everyone was sure this was a major new discovery but Harry refused to sell. Nonetheless, he gave his son, Jerry, an overriding royalty with which Jerry bought his wife, Barbara, a $500 ruby ring. In a little more than a month, the well quit, never to produce again. It proved to be some sort of isolated anomaly and that was the end of the Hollister venture. Jed says that every time this story would come up, his mother would proclaim proudly that she still had the ring.

Harry's son, Jerry, was born in 1927 and proved to be the proverbial 'chip off the old block.' After serving in the Navy at the end of World War II, he managed to break his father's record by getting kicked out of Berkeley not once, but twice. He then joined the family business in 1948 and began a long and successful career, eventually becoming president of Trico Oil & Gas.

Trico Oil & Gas became Trico Industries in the mid 1960s and began manufacturing tanks and pumps. By the late 1960s, the board wanted out of drilling and production so they could focus on manufacturing. Jerry lobbied hard against this shift, but was outvoted.

"Dad said, 'Well, if you're not going to support me as president, then I resign. And since you're going to get rid of these properties, I'll buy out your interest in our jointly owned properties,'" Jed explains.

Many of the producing leases were owned jointly by Trico and Harry individually. Without board support to continue drilling and production, Jerry bought out Trico's interests in several of the jointly owned leases and started his own business called Emjayco.

Business was tough for the new venture for a while, but after a while it began to prosper, thanks in part to Jerry's aversion to borrowing money. "Dad preached, 'Don't borrow money,' so we were able to survive those price cycles because we didn't have any debt," Jed explains.

Jerry died in 1995 but Emjayco continued to operate under his wife, Barbara. When she died in 2006, Jed and his two sisters—Susan Brandt and Ann Magee—reorganized Emjayco as a limited partnership in 2008.

Emjayco, L.P. remains one hundred percent family-owned and is now a fourth-generation company. There are four administrative employees, two of whom are fourth generation employees. In addition, the company employs two full-time and two part-time employees in the field. One of the part-timers, Joe Penny, is eighty years old and has worked for Trico, Emjayco, and Emjayco, L.P. for fifty-one years—longer than any single family member.

Drawing on the founder's interest in real estate, the Magee family has always been involved in ventures other than oil. "My granddad always believed that buying property was the way to go, so we've always been involved not only in oil but in farming, raising cattle and other real estate ventures." Jed explains.

VENOCO, INC.

Above and below: Tim Marquez.

Right: Tar pits.

Opposite, clockwise starting at the top:

Seals floating near the Holly Platform.

Aerial view of the Carpinteria Gas Plant.

Ellwood Field Piers.

North and west side views of the Holly Platform.

In 1992, Tim Marquez founded Venoco, Inc., an independent energy company, with only $3,000 cash and a credit card. Since those humble beginnings, the company has been very successful in acquiring, exploring and developing oil and natural gas properties both onshore and offshore in Southern California.

A native of Denver and a graduate of Colorado School of Mines, Marquez's strategy from the beginning was to grow the company by acquiring undervalued and under-exploited fields and reinvigorating them with the latest technical and operating expertise.

The company capitalized on the low oil price environment of the mid-to-late 1990s by aggressively buying underappreciated and poorly utilized fields and increasing their value through cost reductions and low cost workover operations during periods of low prices, and drilling for new reserves as prices rose.

Venoco purchased its first field—a portion of the Whittier Field—in 1994, as well as the Santa Clara Avenue Field in Oxnard. As growth continued, the company purchased the Beverly Hills Field in 1995 and Mobil's assets in the Sacramento Basin in 1996. Between 1997 and 1999, the company built its offshore asset base by purchasing Mobil's platform Holly and the Chevron platforms Gail and Grace. In 2007, Venoco purchased Texcal with assets in both the Sacramento Basin and Texas.

Venoco was caught up in the Enron scandal that rocked the financial world. Venoco had unwittingly become one of Enron's infamous "off balance sheet" ventures and when the firm began to unravel, Enron executives attempted

to extract money from all their investments and Venoco was impacted.

In 2004, Marquez bought out Enron and its partners, who had paired up with Enron to temporarily gain control of the company, and acquired 100 percent of the company. Oil prices then shot up and Venoco was once again a hot property. Enron, of course, collapsed in disgrace.

Venoco's corporate offices are located in Denver, but the company was founded in Carpinteria and the western region has always been headquartered in the area. Currently, the company operates from beautiful offices overlooking the Pacific Ocean in Carpinteria. Venoco has maintained deep roots along the Central Coast and became the top taxpayer in Santa Barbara County in the mid-2010s.

Venoco currently employs more than 200 people and the company's hiring practices have always focused on finding high caliber individuals. Operational excellence and safety are ingrained in the company culture and Venoco has won many awards for operational excellence both on-shore and offshore in California, where regulations are the most stringent in the nation.

In 2014, Venoco paid more than $24 million in royalties to the State of California, following more than $33 million in royalties in 2013. In addition to generating high-paying jobs at Venoco locations in California and Colorado, jobs are created in sectors that provide services to the industry, including engineering, environmental consultants, transportation firms, and material manufacturers.

Venoco's economic impact is widespread and directly benefits local communities where it operates. In Santa Barbara and Ventura

Counties alone, Venoco paid more than $7.2 million in property taxes in 2013.

Venoco's extensive corporate giving program—known as the Venoco Community Partnership—seeks to reinvest in the company's areas of operation. The grant making program utilizes an online submission system and funding cycles targeting three primary areas: health, critical human services, and education and youth-serving programs.

Employees representing various departments reviews applications and performs site visits to better understand the needs of the community. The list of organizations that have benefitted from the Venoco Community Partnership grants is extensive and includes nonprofit organizations focused on providing services, programs and assistance to under-served and at-risk populations that have broad and significant impacts on the greatest number of people in the community.

Through the partnership, the company strives to be a leader in corporate giving and volunteerism, creating partnerships with organizations that strengthen the communities where Venoco operates and encourages employee engagement.

To date, the Venoco Community Partnership has contributed more than $11 million to nonprofit organizations that support the communities where the company operates and were its employees live.

Marquez is the son of two teachers and a product of the Denver, Colorado, public education system. He credits education for his successful ascent from humble beginnings. In return, Marquez and his wife, Bernadette, have donated $50 million to the Denver Scholarship Foundation to assist graduates of Denver's public high schools attain their ultimate educational goals. They also contributed $50 million to fund the Timothy and Bernadette Marquez Foundation and have personally contributed additional millions to various projects in California, Colorado and Michigan.

Looking to the future, Venoco intends to continue its growth through the acquisition of accretive oil weighted fields and leveraging technical and operational excellence to increase value.

GENERAL PRODUCTION SERVICE

General Production Service is a one-stop shop for customers looking for everything from well site prep for drilling rigs to completion of the well and the infrastructure of the lease.

GPS has survived nearly fifty years in the oil industry because of one thing—the company's diligence and commitment to safety and ability to change its practices depending on the wants and needs of its customers. This strong commitment to change has allowed GPS to be successful in a highly competitive industry.

General Production Service' Rig One.

The company was founded by Charlie Beard in March 1967. At first the company was known as Beard Production, but the name was changed to General Production Service (GPS) a year later.

Charlie began his well servicing career in the early 1960s, working for Cotton Winds and Valley Well Service for six years and for Gordon Holmes for two years. Charlie always had the passion and drive to own his own business and was receptive when he was approached by a superintendent from Gulf Oil Company who asked if he would be interested in getting a rig that would later be put to work for Gulf. Rig One was built just south of Taft on historic 25 Hill and operated by Charlie. Three more rigs were added later, laying the foundation for an up and coming well servicing business.

Charlie's business partner in the early days was Bob Clark, who died in 1969. Others instrumental in the establishment and early growth of the company were Tobe Herndon of Gulf Oil Company and John Hollinsworth of Lloyd Drilling.

Gulf Oil merged with Chevron Oil Company in the spring of 1985 and there was no disruption to the operation; by this time GPS was working for Chevron.

The issues and challenges faced by those servicing oil wells in the valley have changed tremendously since GPS was founded. Advances in technology and equipment, and a greater emphasis on safety around well servicing have revolutionized the industry since the 1960s, and GPS has been deeply involved in promoting these innovations and improvements. This dedication to the industry has propelled GPS to becoming a premier company.

GPS is committed to the satisfaction of its customers, and this begins with the safety and welfare of its own employees. GPS provides its employees with ongoing safety training that includes monthly safety meetings and rewards for employees who follow safe practices. GPS also provides its customers with the highest quality equipment operated by employees who are extremely proficient and experienced in the use of the equipment. GPS is also very committed to the quality of its work and the cost saving ideas it provides for its customers.

GPS, which started with one rig and three employees, has grown to more than 500 employees working for Aera Energy, Linn Energy, E&B Natural Resources, McPherson Oil Company, California Resource Company and many other independents around the valley.

The company now has twenty-two doubles and thirty-six singles, along with a construction division that includes heavy equipment, cranes, welding trucks and an aggregate trucking division.

After forty-eight years in the business, Charlie is now semi-retired. In 2011, Charlie made Rusty Risi, a long-time employee, his partner and turned the day-to-say operations over to him. Risi came to GPS from New York in 1979 and learned the business from the ground up. The warm relationship between the two has kept the company growing and the company's family atmosphere alive. At GPS, employees feel they are part of something great and the company could not have enjoyed the success it has had without the dedication of all those working in the office and in the field.

Charlie has played a vital role in the City of Taft since the company's early days. More remarkable than all the jobs he has provided for local residents over the years is his ability to quietly "make things happen." When Charlie sees a public need, he responds to it. He has been a core figure in the community, sustaining the Taft Oildorado Day's celebrations, the Oil Workers Monument, Little League baseball, Camp Condor and the annual Bike Giveaway. Charlie has been a quiet partner in supporting many other projects that needed help. He is an accomplished businessman, community leader and a proud investor in the future of Taft's youth.

The main office of GPS is located at 1333 Kern Street in Taft. Satellite yards are located in Bakersfield, Belridge and Fillmore (all in California). The company has come a long way from its first yard that was located in Derby Acres near Chevron's 15a property.

General Production Service is dedicated to the future and is working to insure that new and innovative ideas continue to foster growth with a positive impact on safety, the environment, and the industry.

For more information about General Production Service, check their website at www.genprod.com.

Left to right: Charlie Beard and Rusty Risi.

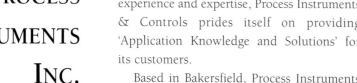

PROCESS INSTRUMENTS INC.

With more than a century of combined experience and expertise, Process Instruments & Controls prides itself on providing 'Application Knowledge and Solutions' for its customers.

Based in Bakersfield, Process Instruments specializes in portable well testing and validation of automated well test systems and provides high quality process instrumentation, valves, and control valves.

The company was founded in 1988 by Paul E. Wade, who started his career in the oil patch by working for his father. Paul's parents, Elvin and Ardith, migrated from Missouri in 1953 and settled on a small oil lease near Fellows, California, where Elvin worked for Chevron. When Shell Oil bought the lease the family lived on, their house was torn down and the family moved to the City of Taft.

"The road to where I am today had a number of challenges and lots of ups-and-downs," explains Wade. "After graduating from Taft High School, it seemed college was not for me, so I started my career working for my dad at General Crude Oil."

Three years later, Wade moved into chemical sales, working for CE Natco Chemical and, later, into the equipment division. "Dave Butt was my manager and he took a big chance by hiring me to sell Natco's oil treating and steam

generating equipment. As I look back now, I realize what a great person Dave was and the opportunity he presented to me," Wade says.

From his job at Natco, Wade moved to ITT Barton Sales where he learned more of the instrumentation and measurement side of the business.

Wade had always wanted to have his own business and, in 1988, decided it was time to start a company representing instrumentation manufacturers. He convinced a number of companies to carve out Central California as a separate territory and founded Process Instruments, Inc. "I started as a one-man operation with a part-time accountant, Carol Bowman, who has continued to stick with me through the good times and bad. Twenty-six years later, she is still putting up with me and is now the controller and operations manager," he remarks.

After struggling for twelve years in the rep business, Wade made a decision to add service to the firm's offerings. "Since much of my past had been spent with supplying instrumentation for Automated Well Testing (AWT) systems, I understood the challenges of maintaining and getting good well test data," he explains. "At the time, no company offered a way to validate the AWT's they get daily well test results from. In 1999, Process Instruments became the first to design, build and commercially offer a service for validating AWT's."

The decision proved to a wise one and, in 2003, the validation system was incorporated into the patented Accuflow Well Test system, creating a state-of-the-art portable well testing and validation system. Working alongside Accuflow, Inc., Process Instruments, Inc., is licensed to use their patented two-phase horizontal well test design. The system is trailer-mounted and PLC based. The company provides one to three-hour well tests (or longer), utilizing proportional to flow isokinetic, positive displacement sampling, static mixers and coriolis mass flow measurement. The results are determined on site and entered into the PLC, where they are stored in a historical file for retrieval or e-mailed to the customer.

"In 2003, we were able to land a contract with Chevron for well testing in their Kern

River Field," Wade says. "Since then, our portable well test and validation business has grown to more than thirty-two systems in California. In 2007 we ventured into the Middle East to work with companies in the country Sultanate of Oman.

"We got our start in Oman validating multi-phase metering systems, which use gamma ray technology for well testing. Due to this opportunity, we have been able to expand into portable well testing and in the Middle East."

Wade explains that the system used in the Middle East is much larger than the one used in the United States. "These units operate 24/7 and are completely self contained," he notes. Currently, Process Instruments serves the countries of Oman and Abu Dhabi in the UAE, where the company employs forty people with additional support from personnel in California.

When the portable well testing side of the business began to grow, Wade brought in two partners—Joe Dravis and Ed Frank—for the sales side. Drawing on their expertise in the control valve side of the business, Process Instrumentation & Controls, LLC was created in 2003. The company's customer base includes oil production and refining, mining and food and beverage. Process Instruments represent much major manufacturers as GE Masoneilan, Rotork, Forum PBV, Vanessa, Keystone, KTM, Truck Process Wiring, Jogler, and Moore Industries.

"With this partnership and by adding valves to our offerings, we have become the premier manufacturer's rep in Central California," Wade notes. "Based on our motto of 'Application Knowledge and Solutions', we have become a key provider to the major oil companies in Kern County, such as Chevron, Aera Energy, CRC and LINN Energy.

To accommodate growth on both sides of the business, the company moved to a new facility in Bakersfield in 2008. The custom-built office and warehouse is located at 8802 Scobee Street. The 16,500 square foot building includes 6,000 square feet of office space with conference and training rooms, a valve automation shop, and warehouse space for valve and instrumentation inventory. Because of the continued growth of the well testing service, the firm has outgrown the current facility and is searching for additional space.

The company employs a total of ninety-five people worldwide.

"Although the price of oil has dropped recently, we believe Bakersfield is still a great place to live and do business," Wade remarks. "We look forward to continued growth and the opportunities that lie ahead. We have been blessed with great customers and employees. It has been a great, bumpy ride. I'm happy to have been able to hold on."

ROYA RESOURCES, LLC

Above: F. Lynn Blystone, president of Roya Resources LLC. Roya is a Persian word for beautiful dream.

Right: Left to right, petroleum engineer, Joseph R. Kandle and registered geologist, Paul D. Hacker looking at a stunning seismic line.

Roya Resources, LLC, is a one-stop shop for generating, leasing of minerals and surface, permitting, drilling, producing, operating and accounting, including revenue distribution of California oil and gas prospects.

Roya, which is a Persian word for 'beautiful dream,' is a start-up firm organized by a group of retired industry executives and investors to pursue new oil and gas opportunities.

producing properties and exploration prospects in these venues. The great central valley of California is the primary focus of the Roya team, which believes that billions of recoverable barrels of oil are still to be found in an area that has already given up more than 15 billion barrels and still produces more oil per day than Alaska's Prudhoe Bay, America's largest oilfield.

Roya's highly qualified core staff boasts a total of 144 years experience. President F. Lynn Blystone (41 years), registered petroleum/drilling engineer Joseph Kandle (51 years), registered petroleum geologist Paul D. Hacker (42 years) and internal accountant Jerold L. Anderson (10 years). Similarly qualified specialists are available as needed.

Roya is familiar with every California onshore basin and can rapidly evaluate both

Roya provides a way for private investors to access industry size plays and obtain significant tax advantages while pursuing extraordinary returns from successful wells and helping maximize tax benefits, even from dry holes.

The firm formed a general partnership, Torrent One, to raise $5 million to drill prospects focused on the east side of the San Joaquin Valley, an area ignored by competitors.

While working for a previous company, the Roya team set the North American record for the longest, fastest single bit run in drilling the 19,086 feet EKHO deep well in Kern County, California. The team has also performed the largest single-stage hydraulic fracturing job (frack) west of the Rockies, putting away 1.3 million pounds of frack sands, and also the deepest frack job west of the Rockies at 18,000 feet. The team is experienced in tar sand (6-8 gravity) to light oil (48 gravity), natural gas and everything in between.

In addition to the vast knowledge and experience offered by the Roya team, it prowls the archives of the California Division of Oil, Gas and Geothermal Resources to spot well histories that indicate wells that were not commercial decades ago, but would be attractive now because of oil prices that are now 25 to 100 times greater. While many such wells are no longer applicable because of real estate development, occasional prospects emerge that still offer exceptional opportunity. Indeed, when oil prices are in the $100 barrel range, every million barrels of proved reserves is worth $100 million future gross revenue (unescalated and undiscounted). Roya has now generated prospects with targets in excess of 200 million recoverable barrels.

Roya Resources, LLC is located at 333 Palmer Drive in Bakersfield. To learn more about the firm, please check their website at www.royaresources.com.

Paul Hacker–Consulting Geologist

This tribute to Bob Hacker was written by his son, Paul Hacker.

Seaman Bob Hacker somewhere in Australia, c. 1943. Notice the rolled up pant cuff for bicycle riding.

Everyone knew him as Bob Hacker, that funny geologist with the hearing aid. But I knew him as 'Pop'.

Pop was born in 1917 and grew up in the Oklahoma dust bowl of the 1930s in the little Alfalfa County town of Carmen. You could not get anymore 'Okie' than that. After graduating from high school he left town with $3.50 and a wicker suitcase and hitchhiked to Houston, Texas. He got a job in the Shell refinery and made his way to blueprint boy, all while attending night school. In 1939 he enrolled at Oklahoma University in Norman, where he followed in his big brother John's footsteps and worked toward an engineering degree. As WWII approached, he got a job as a cement inspector and was involved in the building of the first tin smelting plant in the U.S. and, later, the extension of the runway at Norman Field to accommodate the bigger B-17 bombers.

Pop was a senior when his draft number was called. He tried to join the Seabees and make use of his construction engineering background, but to no avail. However, being a good Boy Scout, he remembered his Morse code and was sent to Moscow, Idaho, to help break and decipher Japanese Code. He was stationed near Melbourne, Australia, for most of the war and later volunteered to go to Guam near the war's end.

While in Australia, Pop began to notice his hearing loss. He was losing the higher frequency sounds but this turned out to be an advantage because he could hear the signal over the background noise that others could not. We would like to thank the U.S. Navy and the VA who always supported his copious hearing aid battery use and paid for his experimental implant surgery, which did not work. Anyone who ever sang with him or asked about his hearing aid would receive the reply 'What?', and soon were as comfortable with his 'handicap' as he was.

After a flight from Guam to Honolulu, then two weeks on an LST with a canvas top, Pop arrived at Treasure Island in San Francisco. While on temporary leave he enrolled at UC Berkeley for the fall semester. At Berkley he took his first geology course from Dr. Robert M. Kleinpell and formally changed majors, receiving his B.A. in Geology in 1948 and his Masters in 1950.

It was at Berkeley, fortunately for me, that Pop met Julienne Hall. They married in July and Pop started working for the Union Oil Company, field mapping in the Santa Cruz Mountains. Field mapping is fun but when you are allergic to poison oak and cannot hear the direction of sounds from a bobcat or a rattlesnake, it can be a little unnerving.

In 1952, Union transferred Pop to Santa Paula California to work with Ed Hall in the old Union Oil building that is now the California Oil Museum. I showed up in 1953. From the time I was quite young he would take me on scouting trips to the rig or to check out some outcrop. We spent many nights in an old Ford Fairlane, waiting for the log run. It was a great time and place to be a kid.

In 1957 we moved to Sherman Oaks and Pop went to work for the Lloyd Corporation. It was a smaller oil company and fit Pop's personality better. Old man Lloyd had grown up on a ranch near the Ventura Anticline. As the story goes, Ralph B. Lloyd was riding up Welden Canyon looking for some cattle that had strayed during a thunderstorm. As he came around a bend in the canyon a gas flare ignited by the lightning from the previous night caused his horse to buck. He was knocked unconscious and when he came to the hill was still burning. Remembering his geology lessons from Berkeley, he leased as much of the east end of the Ventura Avenue anticline as he could and later farmed out the Taylor lease to the Shell Oil Company for a one percent override that dropped after he received $1 million. He used his first million to drill his own wells and completed ninety-one straight producers on the east end of the Ventura Avenue Field.

Pop stayed on retainer with Lloyd through the sale and started his own consulting geology business in 1964. He never hit the 'Big One' but never stopped trying. He may hold the record for the most one and two well discoveries in the U.S. These include Los Posas, Somis and Stone Lake in California, along with Anderson Junction in southwest Utah and Dill Gulch outside Craig, Colorado.

During the downturn in 1985, I joined Pop and started the consulting firm of Hacker and Hacker. We worked together on a number of projects, including trying to help McPherson Oil Company permit and drill the Hermosa Beach Project, a 30 million barrel extension of the Torrance Oil Field.

We worked together until his passing in 1992. He left with a copy of our last well together taped to his hospital wall, "The Waliszek #1 in Lost Hills, 52' of oil sand at 282!" He checked out thinking we had hit the 'big one'.

When I was consulting for Tri-Valley Oil and Gas Company between 1999 and 2010, we were able to realize one of Pop's dreams of extending the Vaca Tar Sands in the Oxnard Oilfield. In 1964, Pop promoted American Petrofina to drill the first successful vertical cyclic steam injection well in the field with the completion of the Vacca-Transamerica #702. In 2007, Tri-Valley Oil and Gas Company drilled and cored 270 feet of Vaca Tar Sand in the Pleasant Valley #1.

With the vision of the CEO Lynn Blystone and the engineering of President Joe Kandle, we drilled the first successful horizontal cyclic steam well in the Vaca Tar Sand. We eventually drilled a total of eight horizontal wells, seven on the Hunsucker lease and one on the Lenox lease. The seven Hunsucker wells have produced over 550,000 barrels to date, or an average of over 78,000 barrels per well. These wells have produced fifty percent more oil than the vertical wells, in one-fifth the time.

I have gone on to have a wonderful forty-four year career in the oil industry and often wonder, "What would Pop do?" In a sense we are still working together. We still believe there is plenty of oil left to be discovered in California. If you are wondering, "Where?" feel free to ask us.

Above: A young Bob Hacker at a Union Oil Company core house, Ventura County, c. 1955.

Below: Left to right, Lee McFarland, (me) Paul Hacker, (Pop) Bob Hacker, and Buzz Fauntleroy. Singing barbershop harmony at the Pacific Section AAPG Convention in Los Angeles, 1986.

WATKINS CONSTRUCTION CO., INC

JOHNSTON VACUUM TANK SERVICE

Top, right: Left to right, Ed and Keith Watkins.

Right: Ed Watkins.

Below: Ed Watkins.

In the mid-1960s, Ed Watkins, Sr., installed the first steam generator for Shell Oil Company near Maricopa. He was working for another contractor at the time, but when he saw what the steam did to the heavy crude, he realized the area was going to boom and that he needed to get in on the action.

Ed, who was born in Oklahoma and moved to Taft, California, in 1952, persuaded his brother, Orvel Watkins, and a friend, Ray Bewley, they should go to work for him in the fall of 1969. "Dad started with just the three of them and one A-frame truck," says Ed's son, Keith Watkins. "What they did—and what we still do as a company—was to put pipe together." He adds that Ed Johnston of Shell Oil was very instrumental in Ed's decision to go into business for himself.

Keith grew up in Venice, moved to Taft, and started working for his dad's company at the age of nineteen, learning the business on-the-job from the ground up.

The company enjoyed steady growth through the 1970s and added five or six trucks and other equipment. Ed, however, was content to keep the company small.

"My dad was the type who believed you paid cash for everything. That was his philosophy," says Keith. This approach began to change, however, when Keith and his brother, 'Butch' began to take over management of the firm in 1983. "We needed more cash flow as the business expanded and we finally persuaded my dad that we needed to acquire some debt to grow. We just couldn't pay cash for every piece of equipment and grow. It was a change in management styles, but from that point on, we began to acquire more equipment and added new customers." Peak employment reached about 150.

Butch left the company in the late 1980s and Keith took over full management responsibilities.

"We want to be successful and we want to grow and for our people to prosper, but we don't have to have a thousand employees," Keith notes. "We want to be able to sleep at night and not worry about debts, so we manage our debt at a level we are comfortable with. Other companies have come in with 500-600 employees, but then an oil slump would hit and they'd be gone. We try to keep in mind that the oil business is cyclical and prepare for that rainy day, because it is going to come.

"What our company does is furnish seventeen ton cranes, welders, backhoes, and other equipment for oilfield construction," Keith explains. "In the '70s my dad and his brother, John Sparks, bought Johnston

Vacuum Truck Company which evolved into a trucking company. Then, in 2002, Watkins Construction acquired Johnston Vacuum Trucking Company."

The company has a number of long-time employees, including Keith's brother, Brad Hazlewood, who is general manager of the construction and maintenance group. Brad has been with the company since the early 1980s. Greg Fanshier, general manager of the trucking group, started as a driver and worked his way up to field supervisor and then general manager. He is in charge of the vacuum trucks that haul water to drilling sites and haul it off. Bud Stubblefield is the mechanic over the construction/maintenance group and Windel Sparks is mechanic over the vacuum trucks.

The current staff includes about 80 employees on the construction side and another 30 to 35 on the trucking side. The company operates from a ten acre yard on East Cedar Street in Taft.

Keith feels the success of Watkins Construction Co. and Johnston Vacuum can be attributed to the quality of its work and the firm's integrity, the ability to respond quickly to our customer's needs, and the safety of our people. "We feel we're the best. Our customers know we're going to give them an honest day's work, that we're going to stand behind our work, and that we're an honest company.

"We've never been laid off by a company or terminated due to poor workmanship. If we do work for a company, we will be there for many years. The only way we're ever going to lose it is because somebody bids lower than we can. We've never been a company that low-balls and slashes wages," Keith adds.

Watkins serves as a board member of the Taft Oil Company and is a former board member at ABC. The company supports local junior rodeos, Taft Oil—Technology Academy Taft College, Future Farmers of America, Links to Life and many others.

Left: Left to right, Greg Fanshier, Windel Sparks, Ed Watkins, Keith Watkins, Brad Hazlewood and Bud Stubblefield.

OIL WELL SERVICE COMPANY

"We can do anything that's needed to be done to an oil well," is the way Oil Well Service Company President Jack Frost describes the services offered by the company he heads.

To be more precise, Oil Well Service is a complete oil field service company, doing workovers, abandonments, completions, geothermal (including shaft driven pumps), electric submersible pumps and water injection wells.

Oil Well Service has provided a full range of services for oil wells for seventy-five years. The company was founded in Long Beach in 1940 by James Sipprelle, Sr., a local banker. He acquired a few partners in the oil business and the company began to grow from small beginnings into a sizable business.

Sipprelle died in the 1950s and his wife, Mary Louise Sipprelle, became CEO of the company. Aided by her associates, she ran the business until their son, Jim Sipprelle, took over. Jim started selling the company to its employees about ten years ago and the company became a 100 percent ESOP company eight years ago. As an ESOP company,

the business is owned by the employees who acquire equity in the firm based on longevity. Frost serves as president and CEO, Richard Laws is vice president, and Mark Smith is purchasing manager.

Frost joined the company as a supervisor in 1995 and Laws began as a rig hand in 1964. Smith started working in the oil fields at the age of seventeen and eventually retired from Occidental Oil. He joined Oil Well Service in 2010.

Oil Well Service now operates forty rigs and employs over 200 people statewide. The company's corporate office is in Santa Fe Springs, where a six acre equipment yard is located. The facility includes five large overhead cranes and a two-story office building that employees consider second to none. Other locations are in Signal Hill, Bakersfield, Shafter, and Santa Paula.

"We are what I consider one of the few Mom and Pop companies," comments Frost. "We treat our employees like family—not like numbers—and we're just one big family. Our people are trained well, they are

experienced, and they work really hard, and because we're an ESOP company, we have very little turnover."

Smith recalls that he was first impressed with Oil Well Service when he was employed by Occidental and relied on the firm to help him keep his rigs operating. "I had several wells that needed servicing," Smith recalls. "So, I called Jack and said, 'I need one—maybe two or three—production rigs with crews. Bring a pump and whatever else you need, and you guys figure out what's wrong with it and fix it.'" Oil Well Service responded to the emergency quickly and efficiently.

"When you have a well pumping seventy-five barrels a day, it's pretty easy to figure out how much money that is when you're making $100 a barrel," Smith continues. "So, speed is essential when a well is broken and needs to be put back in production. The big thing about Oil Well Service is that we are efficient. We go in, fix the oil well, get it back on line and move on out."

Oil Well Service operates mainly in California—all the way from El Centro to Sacramento—but has also done work in Nevada and North Dakota. "We'll go anywhere a customer has a problem," Smith says.

Frost has seen major changes in the industry during his fifty year career and feels it is more difficult to operate today, compared to several years ago. In his view, much of the difficulty is caused by increasingly complicated state and federal environmental control regulations. "California is probably the cleanest state in the U.S., but it has cost us dearly," he feels.

Despite the regulatory frustrations, Frost sees a bright future for Oil Well Service. "The oil business is chicken one day and feathers the next, but we hope to continue to expand, but, it depends on the number of oil wells there are," he says.

Oil Well Service is active in local chambers of commerce and is an active member of AESE (Association of Energy Servicing Contractors). Company employees are involved in a wide range of charitable organizations throughout their communities.

Santa Paula in 1889. This is right after the Adams No. 16 gusher of 1888 and before the founding of Union Oil in 1890.
COURTESY OF CALIFORNIA OIL MUSEUM, SANTA PAULA, CALIFORNIA.

P. WITTE ENTERPRISES, INC.

Right: Philip and Leah Witte at the Oil Barons Ball. Philip was Oil Baron of the Year.

Below: Kenneth Witte graduated from USC Rossier School of Education in 2002.

Much of the exciting history of California oil exploration is lost in the mists of history, but the rich legacy of one fourth generation company is still alive in the memories of Leah Witte. Nearly sixty years ago, Leah helped her husband organize a company that revolutionized the industry with innovative techniques and technological advances. Leah passed away in April 2016 at the age of ninety-three.

Leah was a young schoolteacher in Fellows when she met Phil Witte at a Masonic banquet. A romance blossomed and the two married in 1949. Phil went to work on a lease for the Joseph McDonald Oil Company and all went well until a fateful day in 1960 when he was notified that his services were no longer needed.

Phil saw the setback as an opportunity. He was already in partnership on a gin pole with Vern Dacus and, in 1961, the two borrowed $150,000—an enormous sum of money in those days—and ordered the first brand new Hopper Hoistmobile to appear in the Midway-Sunset Field and went into business for themselves. They called the fledgling company Dacus and Witte Production Services.

"In those days, everyone thought the only way to make a go of it in the oil fields was to buy used equipment but these two ordered, to their specs, a new sixty foot drive-in, three-axle hoist and expected to make a living from it. No one around this area had new rigs and everyone was sure we would go broke, and I must admit it was one tough struggle," Leah wrote in her autobiography, *It's Been Quite a Life.*

Although there were locations in Kern County where oil was literally bubbling out of the ground, Phil and Vern found it difficult to find producers willing to give the new hoist a try. "No one wanted to take a chance on an outfit that had no reputation to back it up," Leah comments.

Finally, a friend who had charge of well maintenance for an independent oil man agreed to give the two young entrepreneurs an opportunity try their rig on his wells. Other jobs followed, and soon other independents were allowing Dacus and Witte Production to have a go at their wells. Meanwhile, hiring employees became a severe problem since no one wanted to work for a company they could not be sure would be around the next day. This meant that Phil and Vern had to do much of the 'grunt' work themselves.

"Before we had the first rig paid off, we borrowed more money to have a single rig built," Leah writes. "Each rig needed a tool truck and a pickup so our fleet grew. Neither of us had ever been so far in debt before. It was scary!"

The tremendous effort required to get the company going took its toll on the men's health and, in 1964, Vern had a major stroke. This meant all the responsibility fell on Phil's shoulders and, before long; he began to suffer from grand mal seizures.

Meanwhile, Leah was teaching school, raising their two sons—Kenneth and Louis—and also serving as the company's only secretary. "I remember doing the books on the back porch of our home on Midway Road in Fellows," she says. She kept in constant touch with the men in the field through a two-way radio.

With all the hard work and long hours, Dacus and Witte Production Service had begun to earn a good reputation for safe working conditions and good equipment. Vern's stroke, however, meant the Wittes had to scrape together enough cash to buy him out.

"Though we were still a hand-to-mouth operation, our CPA advised us to incorporate, so on November 1, 1977, P. Witte Enterprises, Inc. was born," Leah writes. "The new corporation's reputation for excellent equipment and reliable employees began to spread throughout the oil fields, so a move to the big city of Bakersfield seemed the right thing to do."

The yard facilities in Taft were sold and the company rented acreage on Wear Street in Bakersfield. The company remained in that location until it purchased five acres off Fruitvale Avenue and built new facilities.

Kenneth and Louis grew up in the business and started working on company rigs while still in high school. Ken went on to earn a Ph.D. in Education and has enjoyed a long career as an educator. Louis took over operation of the company when Phil retired in 1986.

The firm's reputation was further enhanced when Louis invented and patented the IL2, a piece of drilling equipment that eliminated the need for a derrick man and allowed two men to operate the rig using a joy stick.

In 1998, Louis's daughter, Christy Witte Munn, joined the family business after the birth of her three children; Seth, Caris and Garrett Munn.

In 2001, Key Energy purchased all the assets of the company, grandfathering all the employees and guaranteeing their salaries and benefits. After the purchase, P. Witte Enterprises bought property in Lost Hills that included a couple of producing wells and began developing that property. The existing wells were reworked, drilling of new wells began, and the lease, Salt Creek Oil LLC, now operates seventeen producing wells.

Louis and Christy work together daily and the company's rich history continues to unfold and may someday include the fourth generation.

Clockwise, starting from the left:

Louis Witte checking the pressure on a Salt Creek Oil LLC well in Section 20.

Left to right, Seth, Caris and Garrett Munn.

Christy Witte Munn and Martha Witte, daughters of Louis. Louis is the son of Philip and Leah and brother of Kenneth Witte.

Louis Witte.

SCHLUMBERGER

Above: Logging truck parked in front of Schlumberger Well Surveying Corp. office, Long Beach, California, c. 1940.

Below: Instrument for measuring potential difference, an essential reading for the first electric core drilling operations. This combined potentiometer/galvanometer system included a battery and resistors used to create calibrated voltages.

PHOTOGRAPHS COURTESY OF SCHLUMBERGER.

For nearly ninety years, Schlumberger has provided knowledge, technical innovation and teamwork for the California oil industry. As an international company working in more than eighty-five countries, Schlumberger's real-time technology services and solutions enable customers to translate acquired data into useful information, and then transform this information into knowledge for improved decision making.

Like many enterprises, Schlumberger began as a family business. The world's first well logging company had its origins in the Alsace region on the French-German border where Conrad Schlumberger and his brother, Marcel, grew up.

Conrad and Marcel both wanted to become scientists and were sent to Paris to further their education. Prior to World War I, Conrad, then a professor of physics at the National School of Mines in Paris, developed the theory that various kinds of rocks—sandstone, shale, and limestone—would react differently to electrical changes. By recording the differences, he could learn more about what lay hidden beneath the earth's surface.

When the war ended, Conrad was joined by Marcel, who shared his belief that electrical prospecting could be used to find oil or minerals. This interest in oil exploration brought the brothers to the U.S., where they established an office on the Texas Gulf Coast to test the application of surface prospecting in the oil and mining industries. Their surface prospecting theories proved successful in several Texas oilfields.

As news of the brothers' unique techniques spread, Shell brought Schlumberger to California in 1927 to conduct surface surveys in the San Joaquin Valley. The first electric line log that Schlumberger performed in North America was in Kern County, California, and in 1929 the first electric submersible pump in the U.S. was installed in an oil field near Baldwin Hills.

Schlumberger has provided a number of services for the California oil fields over the years. These technology segments include Wireline, Well Services (cementing, fracturing and sand management), Coiled Tubing, Artificial Lift, Drilling and Measurements, Bits and Drilling Tools, M-I SWACO fluids and environmental solutions, Completions, Testing Services, and Schlumberger Water Services. In addition, Schlumberger Software Integrated Solutions has provided software, petrotechnical, geological and data services.

With 125 research and engineering facilities worldwide, Schlumberger places strong emphasis on developing innovative technology that adds value for its customers. In 2014, Schlumberger invested $1.21 billion in research and development.

Schlumberger employees are committed to working with their customers to create the highest level of added value. The spirit of enterprise passed down from the Schlumberger brothers continues to prevail throughout the company and is reflected in a determination to transform technology into high-quality, cost-effective products and services.

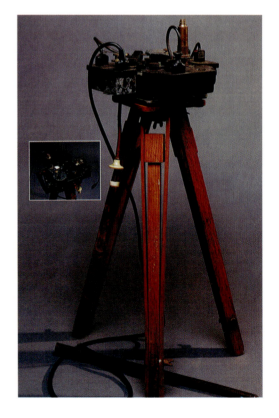

DAVID JANES COMPANY

The David Janes Company, which offers an extensive line of oilfield products, was founded in 1979 by David Janes, his wife, Ruth, and their daughter, Leslie. David and Leslie ran the operation, which was located across the street from its present location in Bakersfield, while Ruth kept the books at home.

David recalls that the oilfields supply industry was dominated in its early days by the steel mills and—since the mills were already selling pipe—they decided to sell piping materials as well.

"They controlled the supply industry in the oilfields through the 1970s, but then things changed very rapidly," David explains. "As foreign competition increased, they began merging and selling off non-performing companies. That's when we opened for business. All of the original steel mills, including US Steel, Republic, Bethlehem and others, are now out of the supply business, leaving the field to companies like ours.

"We started with used materials, which we sold when they were refurbished. A lot of these items were hard to find and we were real busy in the gas country around Bakersfield and in Long Beach," David says.

As an industrial and oilfield supply company, David Janes Company maintains a huge inventory of pipe, valves, welding fittings, flanges, carbon steel, forged steel, and brass and stainless steel pipe fittings. The company also has an extensive supply of 'hard-to-get' items, a huge inventory of tools and threading equipment, safety items and clothing for hazardous work conditions.

As business outgrew the original lot, the company purchased its present yard on Mohawk Street. Forty foot trailers were used for storage in the beginning, but metal buildings were added as the company expanded.

Two of the Janes sons joined the business as it grew. Mathew worked for the company until going into the gun business for himself. Jason now runs the company.

David Janes Company has thrived in the oil, gas, agricultural and industrial supply industry. The company currently has more than 19,000 line items in its 14,000 square foot warehouse. The nearly five acre lot also houses the pipe yard and used surplus materials. There is also a mini-machine shop to do small modifications such as flange facing, bore changes and custom threading. In the 'back lot', customers can find most anything you could imagine, from mobile equipment to oversized valves and piping.

David Janes Company serves most of Kern County and much of the surrounding area. The company currently employs fourteen people, including David, Ruth, and Jason.

For more information about the David Janes Company, check their website at www.janessupply.com.

Above: The David Janes Company is located at 2450 Mohawk Street in Bakersfield.

PCL INDUSTRIAL SERVICES, INC.

PCL Industrial Services, now in its second century of operation, is a diversified, full-service heavy industrial contractor and fabricator servicing a client base in the oil and gas industry that includes petrochemical, mining, power, renewable energy, cogeneration, gas compression, and gas transmission companies. PCL aspires to be the most respected builder in the industry, renowned for excellence, leadership and unsurpassed value.

The company began as E. E. Poole General Contractor, founded in 1906 in the small town of Stoughton in southern Saskatchewan. In 1932, Poole Construction moved its corporate office to Edmonton, Alberta, where headquarters are still located today.

The company entered the United States in 1975 with a project in Colorado Springs, Colorado, and quickly established its U.S. home office in Denver, where it remains today.

In 1977, Bob Stollery, president of the firm at the time, and twenty-four other employees purchased the company from the Poole family, and the current model of employee ownership began. The company name was changed to PCL in 1979.

Although PCL entered the industrial market in the 1940s, PCL did not establish a permanent industrial division until 1979.

The company's history in Bakersfield began in 1981 with the opening of the west coast office of AirPol, Inc., an environmental equipment company specializing in the control of Sulphur emissions from oil fired steam generators. For the next five years, AirPol provided construction management for more than fifty major installations. After 1985 operators were no longer allowed to burn oil in the steam generators, completely ending the need for environmental controls. The firm's six man team wanted to remain in California and relied on established friendships with oil field engineers and began to bid new construction of oilfield facilities.

The ownership changed in 1998 to Fisher-Klosterman, Inc., of Louisville, Kentucky, and again in 2002 to the PCL companies. At its thirty-five acre facility in Bakersfield, PCL operates a code complaint fabrication shop where it performs fabrication, assembly, and field installation of steam generators, pressure vessels, modules, piping, process skid equipment, and custom plate and ductwork.

In addition to new fabrication, PCL does repairs and alterations to existing pressure vessels and steam generators. From upstream oil and gas recovery to midstream gas processing, and downstream to refining and derivatives, this includes site work, foundations and structural steel to equipment and piping systems installation.

The PCL family is now one of the largest general contracting organizations in North America. *Fortune* magazine ranks PCL as Top 100 employer, while *Engineering News-Record* magazine ranks the company among the Top 10 construction firms.

PCL and its employees support a number of community and charitable organizations, including United Way, Habitat for Humanity, one of the top ten teams in Kern County for Relay for Life, Gleaners Food Bank, Corporate Challenge benefitting League of Dreams, Campout Against Cancer, and the Bakersfield Fire Department Toy Drive. PCL is always looking for opportunities to serve its communities.

For more information about PCL, check their website at www.pcl.com.

MIOCENE ENGINEERING SERVICES, INC.

Miocene Engineering Services was organized in 2015 when Alan White purchased the assets of Jack Cook Petroleum Engineering Consultants. Jack has performed petroleum engineering services in California for more than fifty years. Alan retired from Oxy Oil and Gas Company where he served in a variety of engineering, operations management and executive roles. Alan and Jack now operate the business as Miocene Engineering Services, Inc.

Miocene offers a wide range of services, including drilling, completion and workover engineering and management services, production engineering services, enhanced oil recovery (EOR) services, lease operations and operational engineering services, and regulatory permitting.

Headquartered in Bakersfield, Miocene's business is based on its core values of honesty, hard work and trust.

Miocene provides comprehensive drilling, completion and workover services from project inception to oil and gas flowing to sales. Services include well design, cost estimation and Authority for Expenditure preparation, including solicitation and evaluation of bids for rig and ancillary materials and service.

Miocene assists its clients with the evaluation of production performance from existing and new wells, from inflow analysis and well testing. The firm also helps identify bottlenecks in surface gathering lines and process equipment and recommends solutions to reduce back pressure and flow constraints.

Enhanced oil recovery services include design and implementation of EOR projects with a variety of injectants, including steam, water or gas. The firm's engineers also evaluate the vertical and areal conformance of injected fluids and recommend procedures to ensure injected fluid conformance for maximum contact with remaining oil-in-place.

The Miocene staff has more than thirty years of experience in operating oil and gas leases. Whether it is assumption of operations after an acquisition, or operating properties for non-operating investors or financial institutions, Miocene can provide a single source solution for operating oil and gas properties.

When it comes to regulatory services, Miocene provides operational regulatory reporting tasks to insure clients are able to perform the drilling, completion, workover, and operational tasks required to produce and sell oil and gas.

The Miocene staff has extensive experience in all reservoirs and operating environments in California. The reservoirs include extensive experience in drilling, completing and stimulating the Monterey/Antelope Shales in Santa Barbara and Kern Counties, including the Cat Canyon, Elk Hills, Asphalto/Railroad Gap, Monument Junction, Buena Vista, North Shafter, Rose and South Belridge Fields.

Miocene also provides extensive experience in the Stevens sandstone reservoirs at Landslide, Paloma, Yowlumne, Buena Visa, and Elk Hill Fields. Operating environments include urban, agricultural, mountains and sensitive environmental habitat regions of Los Angeles, Ventura, Santa Barbara, Kern, Kings, and Tulare Counties, as well as all Northern California dry gas basins.

To learn more about Miocene, please visit www.miocene-engr.com.

Above: Alan White, petroleum engineer and president, with thirty-five years' experience in executive management, drilling, completions, well servicing, production operations, property acquisition and disposition, reservoir engineering and reserves reporting, and reservoir characterization/development planning.

SULLIVAN OILFIELD SERVICES, INC.

For more than forty years, Sullivan Oilfield Services, Inc., has provided an essential service to the petroleum industry. Sullivan determines the fluid level in an oil well using sonic equipment that provides the most accurate fluid levels and dynograph surveys in the industry.

"The fluid level is something an operator needs to know to determine what they're going to do with a well," owner Jim Sullivan explains. "The operator may be basing a $100,000 job on the information we supply, so it must be accurate."

Jim grew up on a farm south of Lincoln, Nebraska, and was first attracted to California when an uncle, Bob Sullivan, invited him to live with him while attending Ventura Junior College. He then returned to Nebraska and earned a degree in Agricultural Engineering from the University of Nebraska.

Jim was working as a mechanical engineer for a manufacturing firm in Nebraska when he and his wife visited California for a cousin's wedding. "After the wedding, Uncle Bob called me into his office and asked if I would be interested in joining him in a partnership to provide fluid level services," Jim recalls. "At the time it was snowing and below zero in Nebraska, so California sounded pretty good. We started the partnership with no guarantees; if we worked, we ate. If we didn't, we didn't."

Jim and his uncle started the business in 1974 and remained partners until Bob retired in 1982.

Sullivan Oilfield Services has worked from the same location in Ventura—with the same telephone number—for forty years. Jim even uses the same piece of testing equipment he bought in 1974 from Keystone Development Corp., Houston, Texas, and he claims (and his customers agree) it is still more accurate than the latest computer programs.

Sullivan Oilfield Services has enjoyed steady growth over the years, with its fortunes rising and falling with the ups-and-downs of the oil industry, but Jim has kept the company deliberately small. At present, the company serves about twenty regular customers. "We manage to stay busy," Jim comments.

In addition to Jim, the company employs Jeff Hennessey, who has been with the company twelve years, and Bryan Sullivan, Jim's grandson.

Jim is a member of the American Petroleum Institute (API), serves as lector at the San Buenaventura Mission, is past district governor of Toastmasters International, past region director of Serra International, and is chairman of Ventura's 4th of July Street Fair.

Looking to the future, Jim says, "We plan to continue as we are, dedicated to serving our customer with true, timely fluid levels and dynographs."

The company known as Grayfox Oil Co. was formed by Robert E. Long while he was working as a geologist for Union Pacific Railroad in Wilmington, California.

Long organized the company with a partner in early 1964 by purchasing a lease in the Alamitos Heights of Long Beach. The lease contained two idle wells; a drilling contractor had been engaged and all preliminary work had been completed. When Long went on his two week vacation from the railroad, the drilling and deepening of one of the wells from 5,600 feet to 7,800 feet was started. By the end of Long's vacation, the well was completed and put into production.

The well in Long Beach initialed at forty barrels per day and a new career was born. Long quit his job with UPRR and concentrated on his new lease. Within six months, the second idle well on the property was perforated in a zone that had been overlooked in the original drilling. This well came in at 110 barrels per day of oil and substantial gas.

The next major venture for Grayfox came with a town lot lease of 109 acres purchased from one of the major oil companies. Twenty acres of this lease were farmed out to another independent operator to drill four new wells. Those four wells all initialed at about 500 barrels per day (flowing).

The company split up in 1992 after the two partners could not agree on some of the basics of the oil business. Long took his half of the production and continued on as Grayfox Oil Co. This led to several joint ventures with other oil companies. Long ultimately retired and sold all his operated wells, while keeping only the non-operated wells.

Since retiring, Long has continued as a consulting geologist and an expert witness for the Superior Court of California and the United States Federal Court.

DRILLING & PRODUCTION CO. AND E. B. HALL & COMPANY

E. B. Hall & Company was formed in 1936 as a geological and exploration company, with operations throughout the western United States.

Drilling & Production Co. (also known as DRILPRO) was formed about the same time as a contract drilling and lease operating company to serve the needs of E. B.Hall & Company.

The founder of both companies was E. B. Hall, Sr., a geologist who graduated from Stanford University. Much of his early work was focused on the Ventura area where he had been raised; he also did a lot of geological work in the southern end of the San Joaquin Valley on the Tejon Ranch. After entering into lease agreements with the Tejon Ranch, Drilling & Production Co. acquired two portable rigs in order to satisfy the drilling requirements. Whenever the rigs were not busy drilling wells on the Ranch, they would be contracted out to drill wells throughout the state for their own exploration efforts as well as for other companies. Meanwhile, E. B. Hall & Company was also working as the oil operator for Union Pacific Railroad on their Wilmington and Wyoming holdings; this continued until 1969.

After the death of Hall, Sr., in 1949, both companies continued to operate under the management of his son, E. B. Hall, Jr. This was a growth period for the companies, with an increasing number of wells drilled and discoveries made. This lasted until the mid-1970s when it was decided to retire the rigs and concentrate on operations. DRILPRO took over all lease operations (primarily in the South Midway Sunset Field and the Tejon Ranch) and E. B. Hall & Company was dissolved.

In 1982, with the drilling days behind them and the second generation considering retiring, Hall, Jr., convinced his son, Chris Hall, to leave his ten year career in the U.S. Naval Submarine Service to come to work for the company. With oil prices at their peak, there was certainty of prospects to carry the company into the third generation. With the collapse of oil prices in December 1985, the company faced unexpected challenges of continuing to operate with historically low oil prices.

In addition to focusing on the bottom line to keep wells producing, the company was also very active in industry efforts to improve the environment in which they operated. They worked with other companies in oil industry trade association efforts (such as CIPA, IOPA, and the Conservation Committee) on local, state and federal levels. They worked on helping to end the Alaskan North Slope Export ban to reduce the glut of oil on the West Coast; the formation of the Petroleum Technology Transfer Council (PTTC) to help transfer much needed technology to the oil patch; fighting the efforts to impose BTU energy wellhead and severance taxes; and advocating for the independent oil and gas producer whenever and wherever necessary.

One can always say that during the eighty year history of the two companies, business was always conducted with honesty, integrity and ethics; while abiding by the philosophy of "work hard, play hard, and have a good time."

Saratoga Oil Field, Saratoga, California,
unknown date.

CALIFORNIA INDEPENDENT PETROLEUM ASSOCIATION

Above: "CIPA's strength comes from its members who risk their capital every day to fuel our economy. It is our honor to represent them and protect their ability to do business in California," states Rock Zierman, CIPA chief executive officer.

Currently celebrating its fortieth year, the California Independent Petroleum Association (CIPA) is a nonprofit, nonpartisan trade association representing more than 500 independent crude oil and natural gas producers, royalty owners, and service and supply companies operating in California.

The roots of this trade association stretch to the 1930s under groups called the Independent Oil & Gas Producers Association and the California Independent Producers & Royalty Owners Association. The associations merged in 1976 to form CIPA. Since then, the association has kept the political, regulatory, and public policy interests of independent oil and gas producers at the forefront of its agenda.

CIPA's members represent approximately seventy percent of California's total oil production and ninety percent of California's natural gas production.

California has the toughest regulations governing oil and gas production in the nation. Producers must face a tapestry of regulations that are not required in other parts of the nation.

Despite these regulatory hurdles, California accounts for more than ten percent of the oil production in the United States, ranking as the third largest oil producing state in the nation. In fact, when compared nationally, California's largest producing county, Kern, alone ranks fifth behind Texas, Alaska, Louisiana and North Dakota. Californians use 35.6 million gallons of gasoline and 7.2 million gallons of diesel every day. Since there are no interstate pipelines, every barrel we do not produce in California must be tankered or railed in from another country or state.

Out of the thirty-two natural gas producing states nationwide, California ranks thirteenth. More than two-thirds of the state's natural gas production occurs in Northern California and the Central Valley.

In recent years, the industry has been under attack on the regulatory, legislative and judicial fronts by anti-oil activists who want to enact policies to stop all oil production in California. CIPA has been actively engaged at the state, local, federal and legal levels to defend its members from baseless attacks that rely on misinformation rather than the strong scientific evidence demonstrating the economic and environmental benefits of responsible oil production.

CIPA represents the diverse interests of its membership before the California State Legislature, the United States Congress, and numerous federal, state, and local regulatory agencies. The association is an advocate of free market principles, eliminating duplicative regulation, stimulating recovery of domestic resources, and educating the public about industry issues.

Oil production in California benefits the economy as a whole and decreases the state's dependence on imported oil, which is produced with weaker environmental protections. In addition, the industry plays a key role in the state's economy. Oil and gas producers are responsible for nearly a half million direct, indirect and induced jobs that generate $38 billion in total labor income. The oil and gas industry pays more than $21 billion in state and local taxes.

The oil and gas industry is one of the few industries left in California that pays generous salaries to workers from a wide spectrum of educational levels. The average industry salary is approximately $81,000 and one-third of all workers have high school credentials or less.

The industry also boasts an ethnically diverse workforce with more than a quarter of workers from Latino origin, thirteen percent of Asian origin and five percent are African-American. In contrast, 2011 Census data reported that seven percent of Latinos were employed in science, technology, engineering and mathematics (STEM) industries.

CIPA is governed by a forty-eight member board of directors. The board is comprised of producers, both large and small, from the Los Angeles Basin, San Joaquin Valley, Central Coast, and Northern California to ensure the association's policies are broad-based and reflect the interests of the state's producers as a whole.

Above: CIPA served as a major sponsor of the Kern Energy Festival, which celebrated the industry's contributions to the local economy and raised money for scholarships to benefit future careers in the industry.

In 2001, CIPA established the California Natural Gas Producers Association (CNGPA) as a wholly-owned subsidiary. CNGPA was established with the specific intention of increasing public awareness and addressing policy issues specific to the state's natural gas resources. The association is governed by an eleven-member board of directors.

The association has also gotten more involved in campaigns to elect legislators who understand the importance of job creation and meeting the everyday needs of California citizens.

CIPA conducts several events throughout the year, including an annual meeting, which is designed to provide networking and educational opportunities for the association's membership. The event has become the largest annual oil and gas industry event on the west coast. Another key event is the Oil Symposium, which seeks to educate elected officials about the issues impacting production in California. Last year, the event was attended by almost thirty state lawmakers, accounting for nearly one quarter of the California State Assembly and Senate.

As the leading voice for California's independent oil and natural gas producers, CIPA will continue to promote greater understanding and awareness of the unique nature of California's independent oil and natural gas marketplace; highlight the economic contributions made by California independents to local, state and national economies; foster the efficient use of California's petroleum resources; promote a balanced approach to resource development and environmental protection; and improve business conditions for members of our industry.

PETROLEUM CLUB OF BAKERSFIELD

The Petroleum Club of Bakersfield has a rich history with ties not only to the oil industry, but the agricultural, financial, and legal community as well. The club is the gathering place of leaders from young to mature, with a philosophy that reflects a commitment to excellence in all facets. The Petroleum Club is a group that gets things done—with people you can count on.

The club was formed in 1952 by founding board members George L. Bradford, a land-man and real estate developer; geologists William D. 'Bill' Kleinpell, John H. Beach, and Everett W. Pease. Thomas J. Fitzgerald; a geologist and engineer with Gene Reid Drilling, was the club's first president.

The primary purpose of the club was to aid in the association and fellowship of men connected with the petroleum industry and to encourage and sponsor new ideas, which would benefit the oil industry as a whole and provide men possessing special talents with recognition.

The club originally held its meetings at the Bakersfield Inn from 1952 to 1969. The club then moved to the Elk's Lodge at 1600 Thirtieth Street next to the Garces Circle, meeting there from 1970 to 1985. From 1985 to 1993 the club was located downtown in the Bell Tower, a converted church in Old Church Plaza. From 1993 to 2002 the club was located in the old Cask & Cleaver Restaurant on Truxtun Avenue. In 2003 the move was made to the current location on the top floor of the Stockdale Tower, the tallest building in Bakersfield.

"The club truly is a point of light for the city," comments current President Dave Plivelich. "I've lived in Bakersfield since 1981 and the one thing that always amazes me about this community is how connected the people are here, and how down to earth it is."

Prior to the move to the Stockdale Tower, club membership totaled around 200. By 2014, membership had risen to 1,140. Weakness in the oil economy dropped membership to around 950 in 2015, but the club has begun to see an increase because of a wider variety of associations in its membership.

Under the leadership of Plivelich, the club has become the center point for all community leaders and represents not only the oil industry but a wider variety of associations, people and businesses. The Petroleum Club is extremely supportive of local nonprofits and also hosts an annual golf tournament, which provides scholarships for future petroleum engineers. The club hosts many civic service clubs' meetings, including Rotary, Kiwanis, and Petroleum Wives of Bakersfield, as well as serving as a premier location for wedding receptions, class reunions, celebrations and business presentations.

Above: The Stockdale Tower, in Bakersfield, California.

Below: Thomas J. Fitzgerald was the Petroleum Club's first president.

PETRU CORPORATION

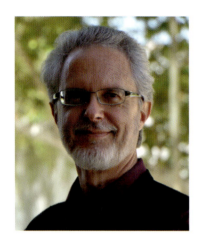

Above: Tim Truwe.

Below: Petru staff meeting, left to right, Marc Arca, Liz Truwe, Tim Truwe, Cathie Feeney and Ken Sammis.

One of those periodic downturns in the petroleum industry resulted in the organization of Petru Corporation, a full-service land company based in Santa Paula. Petru Corporation is a consulting company offering a full line of land services, with a special focus on California and the Pacific Northwest.

During the severe industry downturn in the mid-1980s, Tim Truwe was laid off by an independent oil and gas exploration and production company. Thereafter, Truwe worked as an independent landman and started working out the details and concept of a land consulting and development company. Truwe's experience includes employment in the title insurance industry and the petroleum industry.

Petru Corporation was launched officially in 1986 by Tim Truwe and Gary Peterson. Truwe took full control and ownership of the Corporation around 1991. Although the industry was in recession when the company was founded, Petru stayed busy serving the energy and title insurance industries while expanding its line of professional land services into other areas, including the legal profession and governmental agencies.

Among the early employees of the corporation were Tim's father, George Truwe, who became a landman for the company after retiring as a school teacher and administrator. Other key individuals in the growth of the organization were Tim's daughter, Elizabeth Truwe, who serves as vice president and corporate secretary, and Cathie Feeney, who is executive assistant/landman/bookkeeper.

Petru Corporation has performed title searches, examinations and write-ups for various title insurance companies and their underwriters, for the issuance of title insurance policies on complex multimillion dollar projects. The title insurance industry relies upon Petru's title expertise and has insured their title work. Petru's team of title experts can handle the most complex and involved title issues.

Petru Corporation maintains a large data base of land title records for a majority of the counties in California. This enables Petru to perform title work in a majority of counties without leaving its office, providing clients with cost and time efficiency.

Petru Corporation has been retained to act as an expert witness in litigation involving oil and gas related matters, title, title insurance, water rights and subdivisions of land, and has prepared narratives relative to these issues for court proceedings.

Petru Corporation has provided its services for a wide variety of projects including railroads, water rights, industrial/commercial, natural resources, green energy, rights of way, easements, leasehold estates, agricultural land, subdivisions, mining, offshore petroleum leases and road widening and abandonments. The Corporation has managed title, land and oil/gas projects ranging from as small as one acre of land to more than one million acres.

The Corporation has planned land subdivisions, performed regulatory processing of land subdivisions and lot line adjustments and performed legal lot determinations under the Subdivision Map Act and local ordinances.

Petru performs regulatory (permit) consulting on projects including—but not limited to—oil, gas, mineral, wind, solar and geothermal projects as well as agricultural and business projects. Petru maintains a good rapport with local, state and federal governmental agencies.

The Corporation has acted as an outside or satellite land department for various petroleum and geothermal exploration and production companies, providing title searches, drillsite title reports, lease negotiations, managing land/lease records, joint ventures, acquisitions/divestitures, land availability checks, rights-of-way and permitting. This includes both onshore and offshore on private and federal/state lands.

Petru Corporation has a deep knowledge of natural resource and title matters, they inform and educate their clients in these matters.

Petru Corporation works in conjunction with other industry professions and companies to provide a full complement of services beyond those traditionally offered. Such professional companies include geophysicists, attorneys specializing in oil and gas and real estate, environmental consultants, geologists, title insurance companies, surveyors, engineers and draftsmen.

The Corporation has provided its services to major and independent oil, gas, mineral and geothermal companies, mining companies, solar and wind companies, title insurance companies, attorneys, California Department of Justice, U.S. Department of Justice, local governmental agencies, water purveyors, developers, individual land owners and many others.

The success of Petru Corporation was featured in a recent episode of the *Enterprises* television show, hosted by former pro football quarterback and television personality Terry Bradshaw. Petru Corporation was featured as an expert and authority in the energy field, specifically in relation to oil, gas, mineral, geothermal, green energy, title, regulatory and land matters related thereto. The episode aired on Fox Business Network and various regional networks.

Petru Corporation currently employs eight land and title professionals and is located

at 250 Hallock Drive in Santa Paula and on the Internet at www.petrucorporation.com. The Corporation and its employees support a number of civic and charitable activities, including Boys & Girls Clubs of Santa Clara Valley, Disabled American Veterans, Food Share, Paralyzed Veterans of America, California Oil Museum, several Peace Officers Associations, Smile Train, St. Jude Children's Research Hospital, and the Ventura County Rescue Mission.

The demands placed on the natural resources and real estate industries by the need for use of land and energy for various projects will be a challenge to meet. Petru Corporation has and will continue to meet those challenges by providing its expertise and unique line of land services to the natural resources and real estate industries, professions, businesses, and individuals in need of its services.

Top: Title plant, left to right, Marc Arca and Tim Truwe.

Above: Title plant, Cathie Feeney and Tim Truwe.

NATIONAL ASSOCIATION OF ROYALTY OWNERS–CALIFORNIA

Mineral/royalty owner members of the Hartnell family standing on a dike built to hold oil from Union Oil's great gusher, Hartnell No. 1, also known as Old Maude. This photograph was taken in 1905 on the Hartnell Lease, near the town of Orcutt, Santa Barbara County, California.

NARO-California

Founding Board of Directors:

Edward S. Hazard, CMM, President

Edward S. Renwick, Esq., Vice President

Joyce B. Phillips, Secretary/Treasurer

Stephen P. Kunkel, CPA, CMM

Tiffany M. Phillips

Current Board of Directors:

Edward S. Hazard, CMM, President

Edward S. Renwick, Esq., Vice President

Stephen P. Kunkel, CPA, CMM

Timothy R. Kustic

Roy Reed

Mark O. Phillips

Maribel A. Hernandez, Esq.

Douglas H. Donath

Matthew J. Finnegan, Esq.

The National Association of Royalty Owners (NARO) is the only national organization representing oil and gas royalty owners' interests solely and without compromise. NARO's eleven chapters, including the one in California, educate and advocate for the citizens who own our country's natural resources.

NARO was founded in Ada, Oklahoma, in 1980 with the compelling goal to repeal the Windfall Profits Tax. A group of about forty volunteers joined Jim and Sandra Stafford in this fight. Their successful eight-year battle against the tax is a prime example of what a few people working together can do to make a difference.

Except for some small parts of Canada, the United States is the only country in the world which allows its private citizens to own mineral rights. It is estimated that there are between 8.5 and 12 million private citizen royalty owners in the United States, with over 600,000 in California. The majority of NARO members are females over the age of sixty. They comprise a cross section of the population, including farmers, ranchers, teachers, business owners, blue-collar workers, white-collar workers and retirees. To pay their living expenses, many mineral owners rely on their small royalty income to supplement their social security and/or other retirement income.

Picture the faces of oil around the world. Much of the resource is controlled by not-so-benevolent monarchs, dictators and despots. Not such a pretty picture. Now consider the face of oil in the United States. It is largely the private citizens.

Mineral rights are dominant rights under the law. They are considered private property and as such are constitutionally protected. However, these rights are constantly threatened, especially in recent times. It is the goal of NARO and NARO-California to be vigilant and proactive in the protection of these rights.

There has been a huge increase in domestic oil production in the past few years, with the majority occurring on privately-held property. This has allowed hundreds of thousands of mineral owners to become new income receiving royalty owners. During this same time, production on federally-owned lands has actually decreased. It has been the private sector, and the private ownership of mineral rights, which have spurred tremendous growth in energy production, grown our economy, created jobs and created real wealth.

The California chapter of NARO was launched in 2013. NARO chapters are led by volunteers. Chapters are also located in Colorado, North Dakota, Oklahoma, Texas, Arkansas, Louisiana, New York and Pennsylvania. The Rockies chapter represents members from Montana, Idaho, Wyoming, Nevada, Utah, Arizona and New Mexico. Royalty owners in Ohio, Kentucky, West Virginia and North Carolina comprise the Appalachian chapters.

The goal of NARO-California is to identify, organize, educate, and empower the oil and gas mineral/royalty owners of this state so they will be prepared to manage their mineral assets and to protect their private property rights.

NARO is the only national organization in the United States devoted totally to the royalty (producing sub-surface interests) and mineral owners (owners of non-producing sub-surface rights). NARO is recognized by Congress, state legislatures, national media and the energy industry as the authority on matters pertaining to mineral and royalty owners.

As a nonprofit organization, NARO is funded primarily from member-paid dues and donations.

As an independent organization of mineral and royalty owners, NARO is also dedicated to getting fair play in taxes, legislation and lease dealings and seeks to 'speak the facts' with a strong, unified voice for royalty owners in both producing and non-producing states.

"There is no question that the past two years have been extremely difficult for the oil and gas industry, both financially and politically," notes Ed Hazard, president of NARO-California. "With 2016 being an election year, it is now more important than ever for mineral/royalty owners to educate themselves so that they may protect their assets. It is not only important to be educated, it is also important to actively protect and promote your property rights. This is especially true for mineral/royalty owners in California."

Hazard also says, "It is important to realize that just about everything that impacts the oil and gas producers and the industry at large also impacts mineral/royalty owners. Among other things, those impacts can result from legislation, regulation, taxation, market forces, and the general economy. With this in mind, NARO-California continues in its efforts to build positive relationships with oil and gas producers, legislators, regulators, the media and the public. The primary goal is to protect and promote the interest of mineral/royalty owners by being proactive in the various processes that affect the industry."

NARO also provides its members with the opportunity to enroll in the Certified Mineral Managers (CMM) Program. The CMM Program is an educational self-study course offered exclusively through NARO designed to increase your knowledge of minerals management. To complete the program, participants must accumulate education credits and pass three exams affiliated with their selected level of the program. The CMM Program is administered by the National Royalty Owner Institute (NROI) of the NARO Foundation. All educational seminars and related continuing education courses are professionally designed by directors of the Institute, the Association and the CMM Certification Committee.

NARO-California also conducts free Fuel Yourself With Knowledge educational events in California. These events are designed to educate and inform California mineral owners on how to better manage their assets. These events present a brief overview of the Mineral Management 101 course, the CMM Program, estate planning, tax planning, title curative issues, state and federal regulation. Current local, state and national issues affecting the industry and mineral owners are also discussed. These include pending legislation, regulation and taxation. Oil companies, industry organizations, industry professionals, and the California Division of Oil, Gas and Geothermal Resources (DOGGR) have been valuable participants in these events.

Since its founding two and a half years ago, NARO-California and its members have been actively involved in defeating or modifying for the benefit of mineral/royalty owners a number of local and statewide anti-oil initiatives. These include Measure P in Santa Barbara County, Senate Bill 32, Senate Bill 350 and many other state and local issues. In addition, we have supported numerous industry programs and projects, which would benefit mineral owners and the citizens of this state.

To become a member of NARO-California, or to learn more about the organization, contact NARO at 800-558-0557.

Left: On August 14, 2014, Pat Abel, deputy overseeing regulatory compliance in District 3 for the California Department of Conservation, Division of Oil, Gas and Geothermal Resources (DOGGR) spoke to 100 mineral owners at a NARO-California Fuel Yourself With Knowledge educational event in Santa Maria, California. District Deputy Abel gave a very informative and well received presentation titled "State Regulation of Oil and Gas Operations." She also demonstrated how to use the division's website to access production information. That website is www.conservation.ca.gov/dog.

Right: Eighty-eight year old mineral/royalty owner Winnie Hazard visiting the drilling crew on one of the leases in which she owns an interest. This lease is in the Cat Canyon Field of Santa Barbara County, California. It is operated by PetroRock and Vaquero Energy. Hazard and her late husband, John, acquired an interest in this lease in 1955. This photo was taken in December 2014.

EHRLICH•PLEDGER LAW, LLP

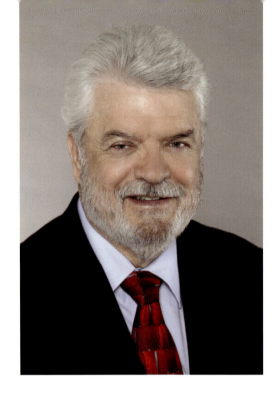

Known as experienced attorneys who get results, Ehrlich•Pledger Law, LLP, is a Bakersfield law firm dedicated to serving the legal needs of oil and gas producers and mineral rights owners throughout California.

Ehrlich•Pledger Law is one of the few independent oil and gas law firms in the Kern County area and the firm offers advice and counsel for its clients' legal matters, both immediate and long-term. The firm has earned a reputation for honest, ethical, experienced representation and has a stellar track record of success both in and out of the courtroom.

The firm's attorneys—Melvin L. "Mel" Ehrlich and Jean M. Pledger—have more than fifty years combined experience and understand the complex matters pertaining to oil and gas law and mineral rights ownership. The two organized Ehrlich•Pledger Law in 2011.

Mel, an experienced oil and gas attorney, has more than thirty-five years of experience, including fourteen years of hands-on experience with two major oil companies. He was a staff attorney for three different divisions of Exxon, and the senior division attorney for Tenneco Oil Company's Pacific Coast Division in Bakersfield. He is well-versed in a variety of oil and gas transactions common to most oil companies.

In 1993, Mel joined Klein, DeNatale, Goldner, Cooper, Rosenlieb & Kimball, LLP, where he became a partner in 1998. He became Of Counsel to the firm from 2002 to 2011 when he and Jean opened Ehrlich• Pledger Law, LLP.

Mel received his B.A. degree from the University of Kansas, where he also earned a law degree in 1974. Except for two years in private practice in Fresno, Mel has lived and worked in Kern County since 1980.

Jean received a Bachelor of Science degree magna cum laude from California State University, Los Angeles, in 1996 and earned her law degree from the University of California, Hastings College of the Law in 1999. She was admitted to the California Bar that same year. Prior to opening Ehrlich •Pledger Law, she was a partner with Klein, DeNatale, Goldner, Cooper, Rosenlieb & Kimball, LLP.

Jean has held a variety of board positions in organizations such as the California Women Lawyers, Kern County Women Lawyers Association, and Kern County Women Lawyers Foundation, including holding the position of president in all three. Jean is an arbitrator for the Better Business Bureau and a pro tem judge for the Superior Court of Kern County.

By pooling their resources, Mel and Jean are able to successfully meet their clients' needs and goals, both transactional and litigation. As a small firm, Ehrlich•Pledger Law provides clients the opportunity to work with an experienced attorney throughout their cases. Mel and Jean take a personalized approach to each case, promptly addressing questions and concerns and working diligently to satisfy the clients' legal needs.

WZI, INC.

For nearly thirty years, WZI, Inc., a specialized environmental and engineering consulting firm, has provided clients a unique level of expertise, an innovative technical approach, and disciplined project management.

WZI offers professional and technical expertise in the environmental and petroleum engineering fields with services including strategic planning, permitting, regulatory compliance, pollution evaluation and cleanups, site assessment, air permitting, health risk assessment and reserve evaluations.

WZI is located in Bakersfield and is headed by President and CEO Mary Jane Wilson and Vice President Jesse D. Frederick.

Wilson, the first woman to graduate from Stanford University with a B.S. in petroleum engineering, is also a registered environmental assessor in the State of California. Recognized internationally as an environmental specialist, Wilson was appointed by the U.S. Congress as the environmental industry expert to review the operations at the Naval Petroleum Reserve No. 1 at Elk Hills. A distinguished lecturer for the Society of Petroleum Engineers, she is a recipient of the society's highest award for environmental achievement.

Frederick has amassed thirty years of professional experience as an engineer, project manager and department manager, having served as the manager of environmental affairs for a Fortune 100 company. Frederick has developed cost effective permitting and compliance strategies for various sized projects ranging from $50 million to $500 million, including international projects, and has been involved in a number of CEQA and NEPA documents. He has been consulted in regulatory development and the public review process throughout the United States.

The firm was founded in 1986, when a new regulatory paradigm was emerging in which oil and gas professionals were expected to incorporate environmental stewardship in their strategic and day-to-day planning. Professional firms supplying traditional petroleum engineering services were asked to help producers in this new effort. WZI, Inc., was founded as a logical extension of the services being performed by the firm Evans, Carey and Crozier. Oil and gas industry clients requested that the firm establish a separate entity capable of focusing on the then

rapidly emerging field of environmental management. In response to the request Wilson, a partner of Evans, Carey and Crozier, along with Casper Zublin created a new firm named Wilson Zublin Incorporated. Within a year, Wilson became the sole principal and she eventually changed the name to the now familiar WZI, Inc.

Wilson, who was third generation in oil and gas was fortunate to be able to draw on her families long ties to the industry and gained board level support from Frank Rosenlieb, Bruce Conway, Ken Evans, Ralph Trueblood, Cecil Basenburg and her father, Jim Armstrong.

Since the firm's founding, WZI's client base has expanded to include projects located both throughout the United States and internationally. The variety of high-caliber work for which WZI is noted comes from the successful project management and the affiliation of leading, specialized talent in complex, industry-specific areas. The success of the WZI strategy to highly sensitive regulatory compliance, site characterization and waste management programs is underscored and clearly demonstrated by the rapid completion and permitting/licensing of some of the largest programs undertaken to date.

WZI's team of professionals provides extensive services to both private companies and public agencies including lead development roles, technical studies, environmental reports, contract services and forensic analysis/expert opinions. The firm demonstrates a clear understanding of the increasingly complex regulatory compliance and licensing statutes, and exhibits a strong capability to discretely manage these sensitive programs. WZI's breadth of experience in permitting, its knowledge and ability to overcome obstacles, and its good working relations with the country's leading authorities ensures its clients high caliber performance and reduced permitting times.

While not inclined to publicize their efforts on behalf of clients, WZI, Inc., has been quietly active in many well- and lesser-known projects. WZI was tasked with assisting the Department of Energy to identify the issues surrounding the Elk Hills Naval Petroleum Reserve and helping guide the process to ensure a successful privatization of the same. WZI provided legal support in many legal cases involving Oil and Gas.

For additional information about WZI, Inc., please visit their website at www.wziinc.com.

Top and above: The groundbreaking of the new building. The gentleman in the red shirt is the other owner, Jesse Frederick. The lady in yellow is the founder, Mary Jane Wilson.

HISTORICAL SOCIETY OF LONG BEACH

Long Beach began in the late 1800s as a small seaside resort. By 1910 it was the fastest growing city in the country. During the next fifty years, the city experienced an oil boom, an earthquake, the buildup of the Navy before and during World War II and immense growth in the years following the war.

Local civic leaders realized the city was rapidly changing and that those involved in its early years were passing away and taking with them the collective knowledge of the city's history. In 1962 they formed the Historical Society of Long Beach (HSLB) to ensure the stories of the city's past would be gathered, historical materials would be collected and preserved and all would be made available to the public. In its early years, the HSLB was managed by volunteers and housed its collection in any available space. Today the HSLB has a staff of three and since 2007 is housed in a city-owned historical building in the Bixby Knolls Neighborhood.

The HSLB has built an impressive collection of photographs, maps, city directories, oral histories, documents, publications, and a newspaper collection that spans nearly 120 years. Also included in the archives are the city manager's files from 1922 to 1952, years that included a major earthquake, the discovery of oil and the development of the city's airport, shipyard and Navy Station. The HSLB makes a point to ensure its collection is inclusive of all aspects of the city's history, such as its Cambodian, African American, Latino, Hmong and LGBT communities.

The HSLB is located on a busy commercial street in a highly walkable neighborhood. The exhibit space opens to the street and includes two galleries that occupy the ground level. Past exhibitions include the city's WPA artwork, the history of its LGBT community, the 100th anniversary of the Day Nursery, the 85th anniversary of the Long Beach Playhouse and Port Town, and the story of the Port of Long Beach. In 2016, an exhibition is planned to commemorate Pearl Harbor and the profound effect it had on the city.

In the research center, the public can examine materials from the HSLB archives.

The archives are used by the HSLB to write the scripts for its signature event, the Historical Cemetery Tour. The popular event features costumed actors who tell the stories of the people buried in the city's two oldest cemeteries.

The Historical Society of Long Beach is located at 4260 Atlantic Avenue, Long Beach, California, 90807. For more information, please call us at (562) 424-2220 or visit our website at www.hslb.org.

The HSLB: Where History Lives!

In the board room of an impressive two-story Victorian building in downtown Santa Paula, California, nine gentlemen solemnly gathered to sign the incorporation papers for Union Oil Company. This was the beginning of a long history of one of the largest oil companies in California. A Queen Anne building with Italianate influences, the Union Oil Building, constructed in 1890 of exceptional Sespe Brownstone and Santa Paula bricks, was not only the flagship of Santa Paula but also the second largest building in Ventura County. Upstairs was home to the Union Oil Headquarters while downstairs housed the Santa Paula Hardware Company. In 1901, Union Oil moved the emerging headquarters to Los Angeles leaving the Santa Paula location to serve as a district office. The downstairs was rented to various businesses over the years until Union Oil and the Petroleum Production Pioneers decided to open a museum dedicated to the history of oil production in California.

In 1950, Superintendent Clarence Froome of the Ventura Division directed employee Robert Daries to create a museum for Union Oil in the main hall of the building. Ventura District Field Foreman Ben Blanchard was assigned to help locate old oil tools. Display cases were built, photos were hung and murals were painted all in time for the museum to open its doors as the first oil museum in California. Roy Lidamore took charge of building an authentic cable-tool rig where visitors were able to walk around it and move levers as if they were working in the field. This remarkable piece of turn-of-the-century technology served as the centerpiece of the museum for many years until Union Oil started making plans to celebrate a wonderful milestone for this enduring company.

Company President Fred Hartley never approached anything except on a grand scale and when it came to Union Oil's, now Unocal's centennial celebration, he would make it extraordinary. What better way to memorialize the company than to revitalize a museum dedicated to the history of oil in California, especially Unocal's. Plans were started in the 1980s to renovate the downstairs exhibits, move the cable-tool rig to a new building east of the original building and restore the upstairs offices to the Victorian glamour of 1890. The effort was exceptional—$2.5 million was expended to make the exhibits of the highest quality. The celebration began with the opening of the newly renovated museum on March 24, 1990. The dedication ceremony included remarks from the heads of Unocal and Santa Paula's mayor, along with tours conducted highlighting the meticulous restoration. Unfortunately, Hartley was in too poor health to attend but new President Richard Stegemeier delivered a notable address in his absence. Lauded as the most impressive museum of its kind, its legacy continues to this day.

Still the most prominent building in Santa Paula, the California Oil Museum hosts 12,000 visitors a year and over 4,000 students in our education program on petroleum and earth sciences. Visitors are entertained by the fascinating exhibits on petroleum, interact with hands-on displays, watch movies on geology and off-shore drilling, view early uses of natural tar by Native Americans and are educated about the significant role oil plays in our daily lives. Funded by donations from the oil industry and the local community, the museum is dedicated to preserving the history of the hardworking individuals and families whose past efforts are embedded in California's Black Gold.

Above: The Union Oil Company in 1890.

Below: A view of the rig room.

COMMERCIAL GLOBAL INSURANCE SERVICES OF CALIFORNIA, LLC

Above: CGIS is located at 100 Spectrum Center Drive, Irvine, California.

Right: The founding members of CGIS, Bart J. Le Fevre and Kristen Kang.

Commercial Global Insurance Services (CGIS), founded 2008, in Irvine, California, is a highly specialized insurance brokerage and risk management consulting firm, dedicated to California's upstream oil and gas companies.

Many of its clients shaped the history and continuing evolution of California's prolific energy industry. From the Metropolitan Los Angeles Basin, up through Ventura and Kern Counties, CGIS is privileged to represent some of the most influential, independent, oil and gas producers in the State.

California's abundant natural resources, coupled with a unique legislative environment, bring new and significant challenges to the industry. While most individuals perceive insurance as a line item for business, CGIS goes beyond the insurance transaction when serving its clients.

California's oil and gas producers are facing new and significant adversities. Recent changes in legislation have a direct impact on their financial well-being. In addition to educating its clients on these regulatory changes, CGIS is actively lobbying the government on behalf of the industry. CGIS takes a day-to-day role in understanding new legislation, in order to educate clients, and advocate for the industry with state regulators.

The founding members of CGIS, Bart J. Le Fevre and Kristen Kang, have dedicated their entire careers to insurance and risk management services, helping California's oil companies address their environmental and public liability concerns. Both are active members of the California Independent Petroleum Association (CIPA) and Association of Energy Service Companies (AESC).

The firm's entrepreneurial spirit, combined with youthful enthusiasm and key market relationships, have made CGIS the brokerage of choice for many well-respected upstream and midstream companies on the West Coast. CGIS is quickly earning an impressive reputation by being a specialist to the growing energy sector, providing responsive, aggressive and proactive services for the direct benefit of its clients.

Headquartered in Orange County, the company strongly believes in giving back to the community and supports many local youth programs in Southern California. The firm is also active in the Avon Breast Cancer Foundation, the Girl Scouts of Orange County, Mothers Against Drunk Driving and the Wounded Warrior project.

More information is available on the Internet at www.cgisllc.com.

GREGORY IGER'S PHOTOGRAPHIC ART, INC.

DBA IGER STUDIO

In 1971, Greg Iger returned to Bakersfield from a stint in the U.S. Army, then time in Los Angeles as a news photographer for United Press International. Hollywood was home and his career took him to shooting movie set publicity, "red carpet" openings and models' portfolios.

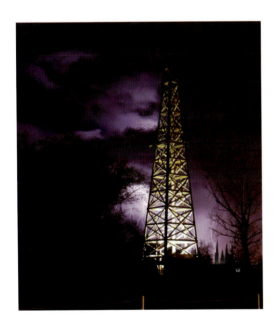

His training was at Brooks Institute of Photography in Santa Barbara. He started "Photographic Art" on a shirttail budget, but soon became one of the best-known photographers in the area. He specialized in commercial photography and created a new look for portraiture in his outdoor studio. Some of his clients were large oil companies like Shell, Aera, Occidental, Chevron, and Tenneco. Other large farming, land and produce corporations like Grimway and Bolthouse Carrots, Tenneco West, Dole and Castle & Cooke were staples for the business, as well as many local entities.

During his many years in Kern County, Iger spent a lot of time honing his landscape photography skills, which spawned two books on Kern County, *Buena Vista—A Pictorial View of Kern County* and *Buena Vista II—Landscapes of Kern County*. Large photographic images for wall décor are now the mainstay of the studio, with clients at offices, hospitals and doctors' lobbies, as well as peoples homes and art galleries.

Iger Studio continues to be a popular favorite for business, personal and family portraits, as well as product, aerial, architectural, construction, oilfield and agricultural photography.

Iger Studio is now located at 211 H Street near downtown Bakersfield. Come in for a visit or call 661-327-2768. Iger Studio is also located on the Internet at www.igerstudio.com.

Above: Greg Iger.

BLACK GOLD IN CALIFORNIA: *The Story of the California Petroleum Industry*

Photographs courtesy of Greg Iger.

TEMPER SCREW.
FIGURE 2

WING
ROPE SOCKET.
FIGURE 1

JARS.
FIGURE 5

SPONSORS

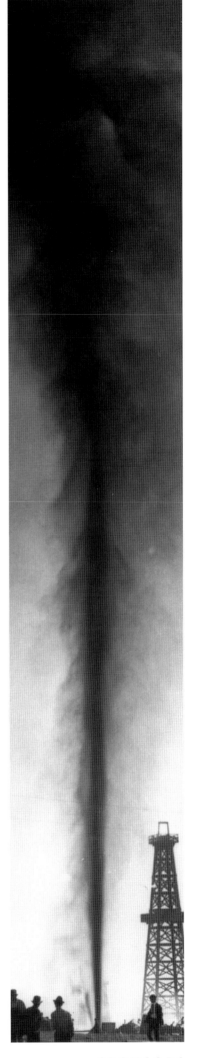

ABOUT THE AUTHOR

ROBERT D. FRANCIS, PH.D.

Robert D. Francis, Ph.D.

Robert D. Francis, Emeritus Professor of Geology and former Chair of the Department of Geological Sciences, California State University Long Beach (CSULB), has thirty-five years' experience in petroleum geology, including industry employment with Getty Oil Company and Texaco, consulting, and university employment. Dr. Francis holds a B.A. in Physics from UC San Diego and a Ph.D. in Earth Science from Scripps Institution of Oceanography; he joined the CSULB faculty in 1987. Specializing in the application of novel geophysical and geochemical technology to marine geology, structural geology, and oil exploration, and especially tectonic evolution of the southern California offshore and nearshore region, Dr. Francis has published over fifty-five articles and abstracts in journals and conference proceedings. Dr. Francis was the initial director (2002-2015) of CSULB's Los Angeles Basin Subsurface Data Center, which has a collection of logs and data files for 30,000 exploratory wells in southern California, as well as legacy seismic reflection images along the southern California coast. Dr. Francis has participated in and organized onshore and offshore seismic data (reflection and refraction) acquisition projects, including in the coastal zone, where specialized techniques are required. Dr. Francis teaches courses in introductory geology, mineralogy, petrology, petroleum geology, geophysics, and marine geology.

ACKNOWLEDGEMENTS

I especially wish to thank my wife, Wilma Marion Francis, for her extraordinary help in finding and organizing photos, tracking down research materials, editing and proofing the manuscript, accompanying me on many scouting trips, and discussing ideas. Thanks to Daphne Fletcher of HPN Books for scouting out many people to donate photos. Thanks to Jeanne Orcutt of the California Oil Museum, and Julie Bartolotto of the Historical Society of Long Beach, for letting me personally pick out photos from their archives. Thanks to Steve Mulqueen for a field trip to the seeps around Santa Paula. Finally, thanks to Rock Zierman of CIPA, John Martini of Freeport-McMoRan, and Frank Komin of CRC, for reading the manuscript and making valuable suggestions for its improvement.